Praise for *What Will Be:*

"*What Will Be* is an engaging and visionary guide to the future, filled with insights on how information technology will transform our lives and our world in the new century."
　　—BILL GATES

"A rich, detailed look at the future texture of our daily lives."
　　—ESTHER DYSON

"Dertouzos delivers a cohesive, deeply thoughtful look at how new gadgetry will change our lives. He also pushes aside all the high-tech hype long enough to explore the potential dangers of this brave new world."
　　—SAN FRANCISCO CHRONICLE

"A mind-expanding roadmap to the coming information revolution."
　　—PUBLISHER'S WEEKLY

"Read Dertouzos's lively and compendious book both for its lucid descriptions of the marketable products that will shape our coming lives and for his passionate expression of the higher moral plane that the Information Marketplace could, repeat could, take us to."
　　—BOSTON GLOBE

"A unique blend of experience, observations and common sense to predict what new products and services info-preneurs will create and how they might change our lives."
　　—WASHINGTON POST

"Michael L. Dertouzos's new book refreshingly attempts a return to basics, eschewing hype and spectacle."
　　—NEW YORK TIMES

"His restrained and readable book makes a good case for optimism."
　　—NEWSWEEK

"Dertouzos maps out the future with the authority of someone who has been within the computer revolution from the beginning."
　　—BUSINESS WEEK

"Highly readable . . . Dertouzos peers into the future and speculates . . . on a lot of things that should matter to businesspeople."
　　—FORBES

ALSO BY MICHAEL L. DERTOUZOS

Made in America: Regaining the Productive Edge

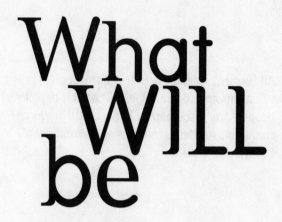

What Will be

HOW THE

NEW WORLD

OF INFORMATION

WILL CHANGE OUR LIVES

Michael L. Dertouzos

Harper*Edge*
An Imprint of HarperSanFrancisco

All futuristic scenarios in this book depict realistic situations and technologies but do not depict real people. Any resemblance between the fictional characters in these scenarios and actual people is purely accidental.

HarperEdge Web Site:
    http://www.harpercollins.com/harperedge
HarperCollins®, ☕ ®, HarperSanFrancisco™, and HarperEdge™ are trademarks of HarperCollins Publishers, Inc.

FIRST HarperCollins PAPERBACK EDITION PUBLISHED IN 1998

Library of Congress Cataloging-in-Publication Data:
Dertouzos, Michael L.
What will be : how the new world of information will change our lives / Michael L. Dertouzos.
ISBN 0-06-251479-2 (cloth)
ISBN 0-06-251540-3 (paper)
1. Information society. 2. Information technology—Social aspects. 3. Information services industry.
I. Title.
HM221.D459   1997   303.48'33—dc21   96-37301

99 00 01 02 ❖ RRDH 10 9 8 7 6 5

To the memory of Leonidas and Rosana
and the happiness of Alexandra and Leonidas

Acknowledgments

Whether they are listed below or not, I am immensely grateful to my colleagues at the MIT Laboratory for Computer Science, whose pioneering contributions, ideas, lively debates, and overall collegiality have shaped much of my thinking and have kept me in love with this field for over three decades. I cannot thank enough all my colleagues at MIT who have taught me so much and widened my horizons and who make this place the marvel that it is. My respect and affection goes, too, to my colleagues around the world who have shaped and continue to shape this exciting field. We have been very privileged to live during a time of profound and exciting change in which we could actively participate.

In particular, I would like to acknowledge Hal Abelson, Eric Grimson, Bob Kahn, Pamela McCorduck, George Metakides, Mike Nelson, Ron Rivest, Bob Solow, Andy van Dam, Steve Ward, Albert Wenger, and Victor Zue, who spent a great deal of their time to help me shape key sections of the book. I thank them from my heart.

Jean-Francois Abramatic, Duane Adams, Bonnie Berger, Tim Berners-Lee, John Seely Brown, Erik Brynjolfsson, Vint Cerf, Dave Clark, Julie Dorsey, Steven Feiner, Henry Fuchs, David Gifford, Chris Halkias, Bert Halstead, Mike Hawley, Barbara Hemmings, Richard Ivanetich, Frans Kaashoek, Alan Kay, Philip Khoury, Isaac Kohane, Dan Kohn, Tom Leighton, Steve Lerman, Richard Lester, Bill Mitchell, Janet Murray, Mike Nash, John Negele, Waring Partridge, Gill Pratt, Stephen Roach, Lisa Rodericks, Ken Salisbury, Jerry Saltzer, Olin Shivers, Dan Stepner, Paul Strassmann, Gerry Sussman, Peter Szolovits, David Tennenhouse, Stephane Tsacas, Barry Vercoe, Albert Vezza, Dave Walden, Larry Weber, and Mark Weiser made invaluable comments, suggestions, and corrections. I am indebted to them for these contributions and for their willingness to help despite their incredibly busy schedules.

My assistant, Anne Wailes, supported by Joei Marshall and Mary Ann Ladd, produced countless copies of the manuscript, fenced with computers, and managed the details of book preparation. I am thankful to them too.

My agent, Ike Williams, and his associate Jill Kneerim helped me frame the book early on, for which I thank them. I also thank Patti Richards, who skillfully publicized the hardcover edition of *What Will Be*. My editor, Eamon Dolan at HarperEdge, deserves special thanks for his unwavering enthusiasm and his impressive ability to make sizable contributions with small strokes of his pen.

Freelance editor and writer Mark Fischetti has my deepest thanks for patiently and creatively working with me to structure language, ideas, and approach for the entire book.

Finally, I would like to thank my family and closest friends for putting up with three years of writing—much of it on their time.

Contents

Foreword

Bill Gates

What Will Be is an engaging and visionary guide to the future, filled
with insights on how information technology will transform our
lives and our world in the new century.

The author, Michael Dertouzos, stands apart from many of the
forecasters and commentators who bombard us daily with images
of this future. For twenty years he has led one of the world's premier research laboratories, whose members and alumni have
brought the world time-shared computers, spreadsheets, the ethernet, RSA encryption, and over forty startup companies.

As a visionary, his predictions have been on the mark: in 1981,
he described the concept of an Information Marketplace as "a
twenty-first-century village marketplace where people and computers buy, sell, and freely exchange information and information services." That's a good description—fifteen years ahead of time—of
the Internet as we know it today.

Not surprisingly, Michael is very much a part of the current action surrounding the Internet: his laboratory's World Wide Web
Consortium involves nearly 250 organizations that help steer the
evolution of the Web.

Naturally, we do not agree on all the specific ways the new world
will evolve or affect us. This is as it should be. There is plenty of
room for new ideas and debate concerning the rich and promising
setting ahead. What's more important is that people become informed, and form their own opinions, about the changes ahead.

When it comes to that future world, what we do agree on far out-
weighs our differences. New businesses will be created and new for-
tunes will be made in the novel areas of activity this book describes.
More important, impending changes in hardware, software, and

infrastructure will alter in ways large and small our social lives, our families, our jobs, our health, our entertainment, our economy, and even the place we see for ourselves in the universe. Whoever takes part in the coming Information Revolution—and that's just about all of us—needs to know *What Will Be*.

Preface

This is a book about tomorrow's Information Age, from the underlying technologies and their uses in nearly every human activity to their social, political, and economic repercussions. One picture pulls together these developments—a twenty-first-century village marketplace, where people and computers buy, sell, and freely exchange information and information services.

The book has three parts: Shaping the Future, which explains the new technologies so that readers can judge unfolding events for themselves; How Your Life Will Change, which imagines how and justifies why our lives will be recast; and Reuniting Technology and Humanity, which assesses the impact of these changes on our society and our humanity. An afterword, new to the paperback edition, presents six action agendas for rich people, for the poor, for business executives and entrepreneurs, technologists, humanists, and national leaders who want to benefit from these changes. Ideas build so that the reader may first gather increasing command over the technology, then its effect on individuals, and then its imprint on society, finally leading to a grand conclusion about the course of technology and humanity in the twenty-first century. The narrative, too, builds around three characters: me and my experiences in part I, you the readers and your lives in part II, and society at large in part III. The book is sprinkled with anecdotes not only for their interest, but also for the lessons they teach about the future.

What Will Be tries to answer questions frequently asked by people not versed in computer technology, examining benefits as well as concerns. It also sets an agenda for information technology and its uses in the twenty-first century. Instead of focusing only on what is exciting, it strives to assess what is real, assaulting along the way the breathless hype presented by the media, which is often haphazard, narrowly focused, uninformed about what is technically possible, and unaware of just how exciting and wondrous the new world truly is. The issues are viewed from both the techie and the humie (humanistic) side, with affection (and occasional scolding) for both extremes. There is ample affection, too, for the users of these new technologies: the ancient humans we all are.

In writing the book I have drawn on industrial and governmental leaders who are shaping the Information Age and with whom I have had the good fortune to work. I have drawn most heavily, however, on the far-reaching research of my colleagues at the MIT Laboratory for Computer Science, which I have had the privilege to direct for over two decades. As a result, the book is skewed toward these familiar surroundings. For this I extend my apologies to my industrial and academic peers around the world whose work is just as pioneering and important and who, together with the people I have known best, created and continue to create the Information Revolution.

In reporting on achievements and on technical matters, the book sticks to facts. But being future oriented, it ventures repeatedly to invention, opinion, and prediction—with the full burden of these excursions resting with the author . . . who had an incredible amount of fun writing it.

I wish that you will also have fun reading it.

Michael L. Dertouzos

PART 1

Shaping the Future

1
Vision

A Home for the Web

The visitors in my office, acquaintances from my native Greece, were touring MIT with their son, who had applied for admission. It was February 1995 in Cambridge, Massachusetts, and the annual ritual of admission was once again under way. The trees outside my ground-floor windows at the MIT Laboratory for Computer Science were dormant, but hopes for college careers were budding.

We were discussing MIT's 150-year tradition of not giving honorary doctorates to anyone, however famous, and many other characteristics of this great institution that made it so attractive to students and faculty alike. Suddenly, my assistant, Anne, appeared in my doorway. "Michael, they need you on the third floor. It's urgent!" I excused myself and rushed out.

I could sense trouble as soon as I got off the elevator. Four members of the team responsible for the World Wide Web—the computer network scheme that had taken the world by storm—were huddled in animated debate over newspapers and e-mail printouts. Two others were on their phones talking with so much artificial calm that it must have been to the press. They briefed me.

It had all started innocently enough the previous day, during a meeting on computer security organized by the Web Consortium, a group at that time of fifty organizations worldwide led by MIT and its European partner INRIA, which strives to push forward the Web standards. At the meeting, chaired by Tim Berners-Lee, inventor of the Web and director of the consortium, a member had asked for a casual show of hands as to which of two proposed security standards the members preferred, based on what they knew so far. Someone had leaked the straw-vote results, and this morning's headlines read: *World Wide Web Consortium decides on Web security standard.* The folks at Netscape, the leading provider of software for navigating the Web, had sent us e-mail threatening to walk out of the consortium because the "chosen" standard was not their favorite. Other consortium members were complaining that they hadn't been consulted. The team was now smoothing their feathers. Albert Vezza, associate director of our lab, was explaining to the reporter who wrote the story why it was wrong; a retraction would be issued the following day. Though I was director of the Laboratory for Computer Science and thus ultimately responsible for the Web Consortium and its activities, there was little for me to do. They were making all the right decisions. I told them so, and urged them to stay cool.

Back in the elevator, I mused that this way of pushing the technological frontier was not exactly what I had envisioned when four decades earlier, as a teenager in the United States Information Service Library of my hometown, Athens, I had come upon the design of a motorized mouse that could find its way through an arbitrary maze. My heart and mind were totally captured by this little machine. I knew that designing mechanical mice at MIT was what I would do for a living. I wasn't aware that the designer of that machine, who would become a colleague, was the celebrated Claude Shannon, who pioneered Information Theory and made the word *bit* something of a celebrity. Nor could I have known that the tiny robot was one of the many crucial advances in a long technical chain that would lead to computers and eventually the World Wide Web.

On this Tuesday almost halfway through the 1990s, we at the MIT Laboratory for Computer Science were still inventing exciting hardware, like bodynets that can link small computerized devices on our eyeglasses and belts with others in our cars and homes, or software that can hold a conversation with a human. But technology had grown to affect the world so profoundly, to become so intertwined with human activity, that it was no longer an isolated pursuit. The rumble blaming technology for the world's ills had long been rising. So it was not surprising to me to have a crisis at the nerve center of the Web that was sociotechnical in nature. Already, in two short years the Web had shed its techie aura and become a major cultural movement involving millions of people. The tens of millions of Web users, from homeowners to CEOs, were growing in number at an alarming rate, adding daily to the cumulative web of information by posting their own "home pages" that described their interests and needs and included writings and other offerings. The (computer) mouse clicks of all these people, like twists on millions of door handles, were opening countless doors to information, fun, adventure, commerce, knowledge, and all kinds of surprises at millions of sites—down the street or a continent away.

Clearly, the new world of information was already affecting everyone's lives. Yet I knew that its present impact paled in comparison to what would be coming in the next several decades. While the media continued to flash old news about information highways, electronic mail, multimedia CD-ROMs, virtual reality, even the Web, newer and more fascinating technologies were already being prototyped in our lab and others around the globe. Meanwhile, the world's economies were getting ready to surrender a huge chunk of themselves to the activities that would stem from these technologies. And the envisioned activities, in turn, were already raising complex new social issues.

It was natural for the media to seize on exciting gadgetry it could already see and understand. But the press was missing much more startling research at labs it never bothered to explore—or that it found "boring" because the technology didn't have adrenal shock value or immediate impact on our lives. On the social and political fronts, too, it was more current to debate pornography on the Internet than the future prospects for war and peace that the Information Age might bring. Mantras like "It's all about interactive TV" and "The medium is the message" were clouding the bigger picture. In a quiet but relentless way, information technology would soon change the world so profoundly that the movement would claim its place in history as a socioeconomic revolution equal in scale and impact to the two industrial revolutions.

Information technology would alter how we work and play, but more important, it would revise deeper aspects of our lives and of humanity: how we receive health care, how our children learn, how the elderly remain connected to society, how governments conduct their affairs, how ethnic groups preserve their heritage, whose voices are heard, even how nations are formed. It would also present serious challenges: poor people might get poorer and sicker; criminals and insurance companies and employers might invade our bank accounts, medical files, and personal correspondence.

Ultimately, the Information Revolution would even bring closer together the polarized views of technologists who worship scientific reason and humanists who worship faith in humanity. Most people had no idea that there was a tidal wave rushing toward them.

I returned to my office and my old friend and his family. They thanked me for my time and left. I would find the son's name on the freshman class list that fall. Good for him; he had won a golden opportunity to see the tidal wave at close range, maybe even make some waves himself.

Our lab became the Web's home through a combination of chance and planning by many people. Three years after inventing the Web, Tim Berners-Lee, still at the CERN Physics Laboratory in Geneva, had begun looking for an institution that would help his brainchild grow. He had been offered opportunities to market the Web by starting or joining a company, thereby entering the club of Internet millionaires. But his idealism, his wish to make the Web a public resource, pushed him to search for a neutral institution. On this side of the Atlantic, as director of a lab that aspired to design the information infrastructures for tomorrow's society, I was looking for a way to bring the lab's celebrated researchers closer to the growing millions of Internet users. We heard of each other's interest and got together. After a dinner in Zurich and a couple of meetings in Boston, we realized we shared the same basic ideas. More important, the chemistry between us seemed right. We felt that we could trust each other.

On February 24, 1994, we clinched the deal. The Web Consortium was planned and formed by Albert Vezza. Tim, who joined MIT and our lab, became its director. Consortium members would pay an annual fee of either $5,000 or $50,000 based on their size. The fee would buy each company and university, large or small, an equal seat around the table where they would debate the future directions of the Web under Tim's leadership and would try to keep it from breaking up into different Web dialects. Within a year giants like AT&T, Microsoft, and Sony had joined, as had innovators like Netscape and Sun Microsystems. By late–1997, the Web Consortium had nearly 250 member organizations.

Fortunately, the consortium members were reasonable about reaching consensus as issues arose. They understood that agreement on standards was crucial to preserving the integrity of the Web across millions of machines throughout the world and, therefore, making money with widely shared Web-based software and services. Sometimes they would summarily endorse a new technology proposed by one member; at other times a member feeling the rush of a recent success might threaten to secede and develop a de facto Web standard on its own. But most of the time members would bend quietly to a negotiated group standard. Other issues, like the desire to control pornography on the Internet, would cause our own Web team to push a standard forward. In 1996 the consortium crafted, in record time, the Parental Internet Content Selection (PICS) standard, which would allow parents to set their computers to block sites based on ratings forwarded by commercial producers, schools, and civic groups of their choice.

The Web, being an international resource, needed political support as well. One night in March 1994, in the mountainous Greek village of Metsovo, I met George Metakides, a friend who had been a fellow advisor to the prime minister of Greece and who, like me, is a dual citizen of the United States and the European Union. George had become head of the European Commission's Program on Information Technology (Esprit), responsible for steering Europe's research in information science. While drinking the local wine and eating delicious Greek sausages, we brainstormed on a dual American-European strategy that would make the Web a truly international standard that served the two homes we both loved. George worked with Martin Bangemann, one of the European Union's seventeen commissioners, who was charged by European President Jacques Delors to develop the European Commission's plan for a Global Information Society. I worked the American side through the Pentagon's Advanced Research Projects Agency (ARPA), the office of Vice President Al Gore, and the President's Science Advisor. The final step was to find a European counterpart to MIT's lab, since CERN, our original partner, decided to stick to particle physics. We selected the French research establishment Institut National de Recherche en Informatique et en Automatique, better known as INRIA. As a result of all these moves, the relevant European and American government officials supported the consortium approach instead of fighting it, extolling it as a model of international cooperation.

Why were so many politicians willing to cooperate so readily? Why have so many companies shown such a keen interest in joining the consortium? Because the World Wide Web is a critical frontier, where some of the major changes the Information Age has in store will begin.

This is also why another even more spectacular scene is under way—the rush of Internet startup companies in the mid–1990s, for which the market has paid billions of dollars, and the equally frantic effort of established companies to exploit the frontier by engaging in mergers, alliances, and wars. Telephone and cable television companies want to control the shipping of information across the giant wire, fiber-optic, and air-wave networks that will reach hundreds of millions of homes and businesses worldwide. Software companies want to supply the programs for the hundreds of millions of computers, televisions, telephones, and other new devices that will communicate over the networks. Computer companies want to supply all the glorious hardware. Media companies that own newspapers, magazines, TV shows, movies, and music recordings want to provide the information content that everyone will be seeking. And as we will see later, a few of the leading companies in each group think they can eclipse the others and provide *everything*—the wires, the software, the hardware, and the programming. The race for supremacy is on.

This drama, while producing a lot of hype, is powerfully real. It is the inevitable consequence of a masterful play choreographed by technical developments in computers and communications in the last half of the twentieth century, and destined to become the socioeconomic movement of the twenty-first century.

Quite a leap from a little mechanical mouse.

The Information Marketplace

The Industrial Revolution began in England when the steam engine was invented in the middle of the eighteenth century. Soon railways and factories appeared, driven by these new mechanized horses. People left the farm for the city, where they would earn higher wages and could buy the improved and ample foodstuffs and clothing the new era provided. They also met with crowded, unsanitary conditions in their new quarters, however, and had to endure the abuses heaped upon laborers, especially women and children.

Technical change had largely stopped by the end of the nineteenth century when a new wave of innovations appeared: the internal combustion engine, electricity, synthetic chemicals, the automobile. The second industrial revolution, as it is often called, made food production possible with many fewer people and improved the production and transportation of goods. Earnings increased, and a new class of white-collar workers emerged. More people became better educated and had money for new services and luxuries. They also became much more mobile.

Again, there was a dark side. Unemployment and social welfare problems appeared along with huge wage imbalances between classes of workers. There was a surge toward materialism and a focus on the self and away from the tightly knit family.

The Information Revolution will trigger a similarly sweeping transformation. The question is what physical and functional forms the Information Revolution will take. What will be its "factories," and what will its people and machines do? I was already trying to envision this almost two decades ago. It was 1980, and I was scripting a talk for a conference entitled Electronic Mail and Message Systems: Technical and Policy Perspectives by the American Federation of Information Processing Systems (AFIPS). Though the personal computer had yet to blossom, I believed firmly that computers would get cheaper and therefore more plentiful. A few years earlier in a *People* magazine interview I had told the reporter that there would be a personal machine in one of every three homes within a decade. Computer networks, I added, would become just as plentiful because of technological developments.

So let me think, I said to myself as I drafted my talk, *what would people and organizations do if they all had computers and all these computers were interconnected?* Stretching, stretching . . . an image flashed before me—the Athens flea market. I knew it well. As a boy I had spent nearly every Sunday in its bustling narrow streets packed with people selling, buying, and trading every conceivable good. I was looking for electronics, especially illegal crystals with which you could build your own small radio station. Almost all of the people were friendly and talkative, tackling every conceivable topic between deals. They formed a community that stretched beyond its commercial underpinnings. There was no central authority anywhere; all the participants controlled their own pursuits. It seemed natural and inevitable to me that the future world of computers and networks would be just like the Athens flea market—only instead of physical goods, the commodities would be information goods.

The "form and function" I hit upon was an "Information Marketplace"—a twenty-first-century village marketplace where people and computers buy, sell, and freely exchange information and information services.

This definition of the Information Marketplace has turned out to be a crisp, simple model that embraces every activity that we might expect or imagine in the new world of information. There is great confusion in the world, today, about what the "Information Age" is, physically and functionally. The model of an Information Marketplace is a clean way to envision both. We will use it throughout this book.

That lucky paper was published in 1981 in a proceedings following the conference. It describes the Information Marketplace in some detail. Here are the opening lines of that vision:

By Information Marketplace I mean the collection of people, computers, communications, software, and services that will be engaged in the intraorganizational and interpersonal informational transactions of the future. These transactions will involve the processing and communication of information under the same economic motives that drive today's traditional marketplace for material goods and services. The Information Marketplace already exists in embryonic form. I expect it to grow at a rapid rate and to affect us as importantly as have the products and processes of the industrial revolution. To sharpen up these abstractions, let us try to imagine the makeup of the Information Marketplace from a point of view that is 20 years ahead:

Large organizations of the year 2000 have been using computers and communications since the late 1980s to communicate business data, electronic memos, and still images among their own plants. Automated inter-organizational transactions have grown substantially in the early 1990s, and the toy personal computers of the early 1980s have become useful and powerful machines owned by small businesses and by many individuals. Office automation has come of age and has led to increased productivity, and to reductions in the use of paper and travel for certain routine activities. A wealth of private and public networks interconnect all of the machines, which number in the ten millions. Entrepreneurs and a new breed of information companies offer a variety of legal, financial, medical, recreational, educational, and governmental information services for a fee. Many traditional ways of doing business have changed. For example, advertising is done in reverse, by a service that responds to consumer inquires with products and services that match. An informational labor force supplies, and many people and organizations consume, all of these services from remote rural or inner city locations.

And here we are! A good deal of this forecast has already happened with the Web and the Internet. I have been telling my story about the Information Marketplace for fifteen years. I continually fine-tune it as technology develops and society reacts to the development. And yet the vision has been consistent in my head, humming along like a well-tuned engine as time goes by. But every time somebody else retells the story, they shape it their way into some contorted vision that may serve an immediate purpose but does not "hum along" in its totality. One of my reasons for writing this book is to tell the whole story, to complete my vision of what will be. The Information Marketplace is already driving a bigger and different transformation than most people imagine. It will change your life. It will change mine.

Let's consider for a moment just one way in which the Information Marketplace model helps us cut through today's hype. The press and most soothsayers tell us that we must prepare ourselves to enter Cyberspace—a gleaming otherworld with new rules and majestic gadgets, full of virtual reality, intelligent agents, multimedia, and much more. Baloney! The Industrial Revolution didn't take us into "Motorspace." It brought the motors into our lives as refrigerators that preserved our food and cars that transported us— creations that served our human needs. Ditto with the new world of information! Yes, there will be new gadgets, which will be fun to use. But the point is that the Information Marketplace will bring useful information technologies into our lives, not propel us into some science fiction universe.

Imagine a salesman in his Louisiana home selling shoes made in Italy to buyers in New York. He talks into his microphone, points to shoe models on a computer screen that the clients also see at their location, and watches their faces on his screen as they ask him questions through their microphone. He does the things he would do in a conventional shoe store, except that he, the salesman, and they, the buyers, are in different places. Doctors, tax accountants, real estate brokers, graphic artists, and a few thousand other types of professionals will be able to engage in their work, at a distance, in a similar way. They won't go into Cyberspace. They'll bring the appropriate new technologies into their professional lives to perform the same work they do now.

The Information Marketplace model helps us straighten out a major related misconception: the world's preoccupation with content instead of work. Common wisdom has it that the information "content" that will flow among the world's interconnected computers will be the text, pictures, audio bits, and video we traditionally call information. But our model also speaks about the delivery of human work over the networks—the salesman's hard work to sell shoes. In today's industrial world economy, activities involving traditional content such as newspapers, books, magazines, radio and TV programs, and Web pages account for about 5 percent of the economy, roughly $1 trillion. However, activities involving information work—primarily office work—account for 50 percent of the world's industrial economy: $9 trillion. Both information content and information work will flow over the Information Marketplace. It's surprising that so many people spend so much time discussing intellectual property rights and payment procedures for traditional content, while they ignore the flow of information work, which will dwarf it.

That's not the whole story, either. Our model indicates that people and their computers will not only buy and sell, but also freely exchange, information and information work. These free exchanges will involve discussions, publications of thoughts and artistic expression, the flow of human help from those who wish to provide it to those who seek it, and much more. Indeed, in the late-1990s on the Web, these noneconomic transactions dominated the economic ones.

We'll explore all this territory. For now, let's agree that the Information Marketplace—not Cyberspace—is the target toward which the Internet and Web are headed.

Back in 1981 I was able to paint a reliable picture of the future Information Marketplace by extrapolating forefront research discoveries to future technical developments, mixing them with ageless human behaviors that never change, and adding some imagination. That is what I intend to do in this book, too. With the benefit of the remarkable progress that has taken place during the years since I wrote my earlier vision, let me now try to provide a few examples of the future that lies a decade or two ahead.

A husband and wife are on vacation in Ruby Creek, a remote settlement in Alaska. Their accommodations at the hostel, which serves tourists and functions as the meeting place, post office, and general store for the sixty year-round residents, are comfortable. The man has not slept well the past few nights, however, finding it a bit difficult to breathe. Tonight he gets worse rapidly. He feels feverish, can't get enough air, and is scared. His wife calls the lone clerk, who helps the man to an eight-foot-high medical kiosk at one end of the hostel's wide lobby. Meanwhile, the settlement's emergency medical technician, whom the clerk has called, arrives. He's not a doctor but knows basic procedures.

The technician asks the man a few questions as he hooks him to several probes on the kiosk. The machine records the man's pulse, blood pressure, temperature, and respiration rate. The technician inserts the man's medical identification card into a slot, and the kiosk sends the data just sensed to the man's primary care physician, who is waking up back in Philadelphia. An alarm sounds on a computer reserved for emergency communications at the physician's home. He finds the data and symptoms disturbing. He transfers the record over a wireless line to Philadelphia General Hospital, where lung specialist Michael Kane can review it. Fortunately he is available.

Using one of the hospital computers, Kane immediately connects to the kiosk. His face and voice appear on the kiosk's small screen. He tells the technician to take an X ray of the man's lungs. By entering a security code, the technician instructs the safety shield on one side of the kiosk to retract, revealing a small X-ray unit mounted on a robot arm.

While the technician maneuvers the unit, the kiosk's computer retrieves Kane's auto-script for handling X rays. The technician only glances at it casually because it is directed at the computers. He reads,

Send chest X ray to A. Smith at Medlab1.
Max transport time 2 minutes.
Min security level is telephonic.
Min overall reliability 99.98 percent.
Return reading to M. Kane at Philadelphia General.

By now the X ray is done and the kiosk, as instructed, sends the image to A. Smith, the hospital's radiologist. Smith examines it and tersely voices his "reading" of the X ray so his verbal assessment will be attached to the record of the image. The three people in Ruby Creek try to relax.

Kane rules out his first guess and asks the technician if the kiosk is equipped with a spirometer and a pulse oximeter. It is. The technician instructs the man to blow into the machine and completes the oximeter test. Kane looks at all the data accumulating in front of him with the wisdom of his specialty and fifteen-year practice. He mumbles to himself, "Your respiration rate is high, oxygen saturation is low and dropping, and the forced expiratory volume after 1 second is abnormally low. No doubt about it: Extremely severe asthma."

Hearing the diagnosis, the man's wife lets out a sigh of relief and tells the physician of their unspoken fear that it might have been worse. Kane is not about to frighten them by telling them, at least right now, that the man might be dead in less than six hours if he does not receive immediate care. Instead he says that the situation is still dangerous and they must move the man immediately to a hospital where he can be watched and, if necessary, intubated. The technician understands more than he lets on. He issues an automated alert for helicopter medevac to Fairbanks General Hospital even before Kane stops speaking. Each hour saved is vital, and the Information Marketplace has saved several hours already. On that frightening night, it has saved the man's life as well.

After 12 years as a loan officer, Julie Cortez is laid off from Rio Sierra Savings in southern Arizona, which was recently taken over by the regional Grande Rio Bank. She applies at the two other banks in her town, but there are no openings. Now what? On the recommendation of a friend, Julie sends a message from her home computer to an employment broker who specializes in financial jobs.

Interested, the agent e-mails back, telling Julie to have her résumé software program contact a special e-mail address. She does so, and a long form on the broker's computer is automatically filled in with details supplied automatically by Julie's program—except for six questions the program did not understand or had no answer for. Julie completes them with her keyboard in a few minutes.

An hour later Julie receives a message that the broker has reviewed a number of openings his job-search software found and thinks there are five positions for which she might qualify. Julie reads the details of each. Three sound intriguing. She agrees to an online interview for each and to the broker's fee of 10 percent of her first three months' salary.

The broker arranges the interviews. For each one, Julie sits at her computer, and she and her potential employers see each other on their screens. She does well and gets two offers, subject to a live interview. One is particularly attractive; if she gets it, she'll be one of seven loan officers for a growing international bank with online banking services throughout North America.

With great relief she learns she'll be able to work from her home with computer equipment that the company will provide. They will also pay for certain modifications to her home work area, subject to her agreement, that they have found to be effective for remote work. She will have to spend one week each quarter at the bank's headquarters in Dallas for live training and networking with her six principal co-workers and headquarters' personnel. She will need face-to-face interaction to get to know them well enough to trust their judgment, as they will confer during video discussions with potential borrowers in the southwestern United States and northern Mexico, her territory.

Julie decides to go after this job. She will miss being part of a living office as she had been before, but this is a good opportunity. She'll make it a point to meet her old co-workers in town for lunch. Besides, she will now be able to see her two teenage sons when they come home after school, a definite plus.

Now let's visit the shopping district in Paderborn, a medium-sized town in Germany. It sports an attractive store called World Shop. It doesn't carry a single product, but it does have thirty-five modest-sized cubicles, each one outfitted with a large video screen, a small tabletop with a keyboard and computer mouse, a few chairs, microphone systems that recognize speech, special goggles, and some neatly instrumented gloves. All but four of the cubicles are occupied.

In one cubby three university students are shopping for dresses. One at a time the women stand in front of a 3-D scanner that takes their measurements. They each put on a pair of goggles and see a lifelike image of themselves in each dress they select from the online catalog, as if they were looking into a mirror. Sometimes they laugh at one another; sometimes they pass the goggles around raving about how "really cool" they look. They each order a favorite, which will arrive at their home the following day . . . and should fit them well.

In the cubicle next to them, a tea lover is searching for exotic blends. He settles on a somewhat predictable Ceylon and a more adventurous Korean brand that he noticed had been highly rated by previous buyers. They will be shipped to his office downtown.

Next to him a man is shopping for a car. He was finally given that big promotion, and his daughter, who is finishing school this year, is going to get the old VW anyway. He wants something racy. With his goggles he can see different models as if they were parked in front of him. The new Mercedes coupé looks enticing. He sits in his chair and puts on the special gloves that help his hands "feel" objects he will see. With a mouse click he finds himself seated inside the coupé. He extends his gloved hands and "grips" the steering wheel. Nice. He then reaches his right hand over his head and suddenly hits the roof. Oh no. He slides his hand toward his right ear, and feels that the roof is grazing his hair. One good bump in the road, and he'll have a good bump on his head. Ah, but the seat can be lowered. Good. It's time for a trip to the local Mercedes dealer for a real test drive.

Yet another patron is selecting art from a huge painting gallery that is highly distributed with suppliers of art spanning the globe. Two colorful large prints have just rolled off the shop's high-resolution printers and are already propped up against the cubicle wall. He now says *Escher* and finds twenty works by the master of geometric deception. The man beams. He never would have found these in the old frame store. And the shop at the art museum in Düsseldorf, one and a half hours away, does not have many of the paintings he cares about. He'll print one Escher, automatically charging the cost to his bank account.

There's no telling what products and services individuals in the other cubicles are shopping for. They all seem to be having fun, even though some of them, like the tea drinker, probably could have found their treasures from their home computer. They prefer the advanced equipment at World Shop, the easy and uniform searching made possible by the store's refined software, and the social encounters that can only be had by hanging out at the mall. They also count on the store's reputation and its flexible return policy.

As these vignettes suggest, information marketplaces can be metropolitan in size and scope, like the health care network. Or they can belong exclusively to one type of organization, like the banking industry. They may have a selective international reach, like World Shop. Or they might operate on a massive scale, linking people at a hundred million computers who might join together for a truly global event. And you, sitting at your family's computer in your quiet living room, could leap from one to the other with a simple spoken command or a mouse click.

Infrastructure Is the Key

With such size and richness, the Information Marketplace is more extensive than a village market. It is closer to a bustling metropolis where many people, shops, offices, and organizations busily conduct millions of personal and commercial interactions in pursuit of their own goals. In a real city, these activities are supported by a shared foundation—an infrastructure of roads for the transportation of people and goods; of pipes and wires for moving water, electricity, and phone conversations; of doors, locks, and police that maintain order; and of some agreed-upon conventions like a common language and accepted behaviors that facilitate interactions among the city's people.

In exactly the same way, the Information Marketplace is built on a shared infrastructure made up of all the information tools and services that enable its many activities to function smoothly and productively. This infrastructure will be distributed and owned by all of us, not a single organization. It will move the data, voice, text, and X-ray images in the severe-asthma scenario by negotiating automatically with phone, cable, satellite, and wireless carriers and with the kiosk and computers at the radiology lab and doctors' offices. The infrastructure will support all the online interviews and reviews Julie Cortez will perform in her daily job. And it will help transact all the business from the World Shop.

None of these scenarios is fully possible today. Despite the wonderment in the press, and the hype in phone and software companies' ads, the Information Marketplace's infrastructure is far from complete. To objectively test whether a true information infrastructure exists anywhere in the world today, let's check what's out there against the key properties of well-known infrastructures—the telephone network, the electric power grid, and the highway system.

The most noticeable property of any infrastructure is its *wide availability:* there is a telephone and an electrical outlet within easy reach in every home and office, and a road is almost always waiting quietly right outside. Infrastructures are also *easy to use:* pick up the phone; plug in an appliance; get into your car and drive away.

Infrastructures are readily *scaleable:* local phone networks, power lines, and roads are connected to form regional phone networks, power grids, and highway systems, which are then connected to create larger national and international infrastructures.

The most powerful property of a true infrastructure is that *it makes possible numerous independent activities.* The telephone infrastructure makes possible millions of conversations every day that span a wealth of different topics, from business transactions to love chats—not to mention fax and modem activity. The electric system powers thousands of different appliances, from steel-melting arc furnaces to kitchen can openers. The highway system allows motorcycles, cars, trucks, and buses to move anyone and anything anywhere.

How do today's computer and communication systems fare against these key properties of traditional infrastructures?

Computers are widely available, at least in the industrially rich nations. Communication services can also be easily purchased, although today's telephone networks cannot transmit data fast enough for some existing and many forthcoming applications; it would take a month to ship a full-length high-definition color movie through an ordinary telephone line!

Information networks are scaleable. The famous Internet, which ties millions of computers and their users, has mushroomed to its current size by linking thousands of smaller computer networks.

We'll say more about the Web and the Internet. Meanwhile, think of the Internet as a postal system for shipping raw information among the world's computers. And think of the Web as a specific way of using this system to view and visit information on distant sites by clicking your mouse.

Unfortunately, computers and communications networks are not easy to use. The manual for a word processing program is as thick as a dictionary. Even telephones have become complicated, not to mention inhuman, like the automated corporate answering systems that force us to suffer through tedious push-button choices before letting us talk to a real person—if at all.

The most important property of an infrastructure—the ability to make possible numerous independent activities—is not met by today's information infrastructures either. Surely, individual computers do support many useful applications, from spreadsheets to computer-aided design. But they cannot perform easily thousands of different tasks over a network. My computer cannot find me the car with the greatest headroom, because different manufacturers keep their data in different forms and on different sites. This is the norm today. Different machines and different software packages use different rules. You must stand on your head and use all kinds of arcane mechanisms to make any sense out of them. Browsers and the Web don't help in this regard, because you end up doing an inordinate amount of work searching without any assurance as to the outcome.

So we must objectively conclude that there is no true information infrastructure anywhere in the world today. As we cast this shockingly negative shadow on today's information systems, we should remember that we are only thirty years into the new technologies of information. It took more than a century to move the world from steam engine to internal combustion engine. Some patience with this young field is in order.

Okay, so there is no true information infrastructure around. Surely someone is building it, right? Yes and no. The Web and the Internet are the right start, but as we shall see they are still a long way from being there. And the large corporate forces—the telephone, media, software, and hardware companies—are not helping. They naturally view the future as a place predominantly for their products and services. The infrastructure that ties all these participants together is of secondary interest to them. They each want to set up their stores along the highway, and they don't see it as their job to actually lay down the pavement. So where will the infrastructure come from? As we'll see in chapter 4, every one of these actors will have no choice but to contribute. The challenge before them, and the rest of us, is to get competitors to work in concert to build an infrastructure, rather than let it happen as it might. This could make a difference of a decade or more in the arrival of an Information Marketplace that provides real utility and opportunity.

Governments around the world have mixed attitudes about building the information infrastructure. Even though past U.S. and European governments helped build their national highway, rail, telephone, and power systems, their contemporaries are largely staying out of the infrastructure business, relying on industry instead. This approach is highly questionable. It is rooted in what is politically fashionable today—deregulation and small government—and not in what is prudent for the common good of each nation's citizenry. Japan and several Asian countries are taking a more centralized approach, building better infrastructures that will give their people a decided advantage. The problems they face are excessive management and control of the new resource, but at least it will be available. Government should focus more intently on creating a truly solid infrastructure for the Information Age. Without it, we will be driving exquisite automobiles on an ill-conceived network of dirt roads.

Let me return to my tantrum about automated answering systems, because it holds a bigger message. Surprisingly, we have arrived without objection, let alone violent revolt, at making phone calls and being ordered around by a mechanical voice that says, "For a list of employee extensions, please press 1. If you want Marketing, press 2. If you want Engineering, press 3. If you . . ." We dutifully execute these orders, precisely as directed, lest we miss a turn. The image we are supposed to carry from this technological advance is that machines are doing our work. Baloney! Civilized humans are expending valuable portions of their lives executing instructions dispensed by a hundred-dollar computer! The companies that use them will claim, "It is cheaper to work this way." Yes, for *them*. Not for *me*. In the long run it may not be cheaper for them either, as customers switch to new services that use visual displays for the menu of choices or, even better, a combined system of humans and machines that is helpful . . . and faster for the customer. This kind of dehumanizing misuse of computers will have to be overcome before the Information Marketplace can be truly beneficial.

Such abuses of modern technology provide fuel for humanist readers, who may gleefully conclude that finally a technologist has admitted that the world's techies are destroying our humanity. I hope that by the end of this book techies and "humies" alike will agree that such extreme positions, on either side, are misguided. In fact, I intend to show that the Information Marketplace will inevitably cause us to bring closer together our humanistic and technical sides, which have been artificially split for centuries. The split runs counter to human nature and prevents us from coping with, let alone making the most of, the increasingly complex world around us.

Global Fever

There is hardly a nation that is not interested in leveraging the potential of the Information Marketplace. The United States is clearly ahead; most of the Web sites and Internet traffic are in the United States, as are the leading information companies and startups. There is political interest at the highest levels too. Not only did Vice President Gore lead others in envisioning a National Information Infrastructure (NII), but, incredibly, in the mid-1990s, he spent on average more than an hour a week with his staff furthering progress in the government's plan. Indeed, Gore uses the tools he preaches about. In preparing for his MIT commencement address speech of June 7, 1996, he used the Web extensively to obtain ideas from some one hundred graduating students about the opportunities and pitfalls they saw ahead.

The NII initiative includes over $1 billion annually in research and development at ARPA and another ten U.S. government agencies. It also encourages the wiring of all schools and libraries and the reduction of regulatory barriers. This last objective culminated in the 1996 Telecommunications Act, which repealed regulations in place since 1934 and equalized the rights of cable and telephone companies to provide telephone, television, and information infrastructure services. This is a boon to competition that will serve people well with more services and lower costs.

On the other side of the Atlantic, the European Commission published in 1994 its Bangemann Report (Commissioner Martin Bangemann headed the effort), which is a plan for the Global Information Society, as the Europeans like to call the Information Marketplace. It is very similar to America's plans, though Europe's implementation is slower than America's, especially on the regulatory liberalization front.

Japan plans to turn its telephone network into the nation's information infrastructure using the so-called ISDN (Integrated Services Digital Network) approach. The intention is to link every home and office with glass fiber by the year 2010. The cost is immense—over $300 billion—but the Japanese will do it. This will make Japan's communications pipes enviably more capable than those of the United States or Europe. However, Japan is not paying as much attention to the rest of the infrastructure: tools and services. What good are twenty-lane highways if you don't have the vehicles to use all this capacity?

The excitement of the Information Marketplace is not limited to large countries. Singapore has declared an aggressive information plan that started with the Information Marketplace ideas its leaders adopted following their visits to our lab in the 1980s. With Prime Minister Goh Chok Tong dedicated to the mission, Singapore is well on its way.

People in Hong Kong, Taiwan, and Australia are beginning to see their role in the emerging global Information Marketplace as brokers who would match suppliers and users of information and information services. The Baltic states of Latvia, Lithuania, and Estonia and other central and eastern European countries look to leapfrog in part Western infrastructures, replacing portions of their unusable telephone networks with advanced satellite and glass fiber telecommunications. China has also begun to tap the Internet. The fabric of the Information Marketplace will expand vastly when its huge population "goes online."

It is hard to believe that at the close of the twentieth century there will be a single country that has not declared its interest in becoming a player in the Information Marketplace. As we shall see, this worldwide involvement will ensure the global reach of information that will improve economies and strengthen democracies but also aggravate tensions and problems arising out of cultural friction.

The sudden realization across the globe that the coming world of information will play a key role in people's lives has caused different nations to put their own imprimatur on the new "thing" I call the Information Marketplace. The race to coin a name that will prevail is yet another indication of just how big everyone expects the "thing" to be.

There's no telling which label will stick. The name National Information Infrastructure (NII) is unfortunately parochial, because information infrastructures cross national boundaries as easily as they cross buildings and city limits. It has recently been replaced by the Global Information Infastructure (GII), a better name. The name Information Superhighway is also limiting, because it focuses on the transportation of information—what the telephone system does today—while the real benefits of the new "thing" derive from moving *and* processing information. The other name we often hear, Cyberspace, is also flawed, as we have already explained. The Europeans have deliberately chosen Global Information Society to emphasize the societal dimensions of the new medium, like learning and health care, and to deemphasize the hardware implied by images such as highways. This name is attractive because of its broad reach, but to the wrong ears it may build loftier expectations about the role of information than its architects intended—perhaps like the *bulldozer society* might have been for the Industrial Age.

Having poked some affectionate fun at these names, I now come to the delicate task of critiquing my own wordchild, Information Marketplace. By sticking to the familiar, human, and lively notion of *marketplace,* I wanted to characterize information as a useful, no-fuss good, closer to physical goods and services than to ideologies, that will be used in our everyday lives. However, to some ears Information Marketplace sounds suspiciously capitalistic and driven by economic goals.

What is most likely is that none of these early names will survive. Do you ever say, "I'll now use the world's telephone communications system," or ". . . the electric power grid," or ". . . the automobile transportation infrastructure"? Certainly not. You say, "I'll call," "I'll plug it in," "I'll take I–95." I'll see you on the Web.

Questions

Trying to foresee the future uses of the Information Marketplace is as futile as Alexander Graham Bell's having dreamed that his invention would lead to answering machines, 900 numbers, phone sex, dial-a-prayer, faxes, and cellular car phones (there was no such thing as a "car"!). Now, as was the case then, we *can* say with some confidence that the new technologies will profoundly affect every corner of our personal and professional lives. What we have trouble discerning is just how it is likely to happen. Nevertheless, that is what I intend to do. Some of what I predict will undoubtedly prove wrong, but I hope to identify some lasting patterns of tomorrow's Information Marketplace, along with its major promises and problems.

Let's begin our journey with a list of questions the book will tackle.

Ideally the Information Revolution will repeat the successes of the Industrial Revolution, except that this time brain work instead of muscle work will be offloaded onto machines. Is this going to happen, or are we going to be handed high-tech shovels, like today's Web, that force our eyes and brains to still do all the shoveling? Are we going to gain greater and faster access to needed information, along with a greater individualization of products and services, or will we drown in info-junk? What should software and hardware vendors of the twenty-first century offer to propel the Information Marketplace beyond its current medieval stage?

Will computers increase the industrial performance of the world's nations, or is the help they offer irrelevant to that quest? What tools are likely to grow on these infrastructures for carrying out electronic commerce and groupwork? What will happen to employment? Will any economic activity remain unaffected?

Will our way of life improve through cheaper, faster, and higher-quality health care and a greater access to knowledge? Or is better information a minor player in these quests? Will the rich who can sooner afford these technologies get richer? Will the poor be given new leverage, or will they just be left further behind?

What new software will flourish in the Information Marketplace? Which programs will be used in everyday life, and which will be reserved for specialized situations? What new gadgetry and interfaces might appear, and how will we use them?

How close to the real world can we get with goggles, tactile bodynets, virtual "feelies" and "smellies"? As greatly enhanced entertainment comes to our living rooms, will we derive greater enjoyment or become lazier couch potatoes? Will we surrender physical human interactions to the artificial virtual-reality cocoon?

Will people benefit by preexperiencing future vacation spots, or will they become jaded by all this foreknowledge and lose the pleasure of discovery and the spontaneity of an adventure into the unknown?

What kinds of battles will be fought as everyone rushes to profit from the place? Who will be the winners and losers? How will the infrastructure look when these battles are over?

Will ordinary citizens be better heard by their governments, or are electronic town halls impossible to achieve? Will our privacy be assured on the world's information marketplaces, or will Big Brother end up knowing more about all of us? Should we change laws to protect ourselves from this new technology? If so, how? How might war and peace be affected?

And what of human relationships? Will they become stronger, rooted in the Information Marketplace's rich global embrace, or will they become more transient and fickle? With all the people of this world only a few mouse clicks away, will a universal culture emerge to bind us together, or will this increase in proximity cause overcrowding, fights with our new neighbors, and the rise of info-predators and info-crime? Ultimately, we want to know what human qualities pass and do not pass through the Information Marketplace; can we love or hate over a computer network, or will these forces of the cave only be possible as we stand face-to-face before our friends and adversaries?

Let's get some answers.

2
The Revolution Unfolds

Birth of the Computer Community

Looking at the ascent of the World Wide Web from nothing to several million users in a couple of years might lull us into thinking that the new world of information is a recent arrival. Not so. Its roots go back thirty years to when the first computer communities were formed. What's happening now to millions of people—to each of us—and what will happen in the future to hundreds of millions is rooted in what happened first to a small number of pioneers. Understanding those origins and the key intervening landmarks is essential to understanding where we are today and, more important, where we are headed. Already the emerging Information Marketplace offers potent proof that people and companies that ignore the lessons of this history are condemned to repeat it—at their peril. And surprisingly, we will find in the past some key developments that have not yet come to fruition, even as millions of people use the Web and the new technologies of information.

I am sure you are eager to get to the whiz-bang stuff, but that is part of the problem. Every day there is another news story about another "amazing" technical development, but to people who know the history, many of the breakthroughs are not, and many may have no relevance whatsoever. As we make our way through the unfolding revolution to the present, we will also see the emerging postures and strategies among the computer, software, media, telecom, and cable companies. We'll conclude the chapter with the battles to come among these giants—which are crucial to the formation of the Information Marketplace.

As I bring you quickly to the present, I will tell it like it is and begin doing what I will do throughout the book—debunk the hype out there, help you see through the haze of opinions, public relations, press stories, and advertising so you'll be able to judge for yourself what is important and what isn't. Once the story is done, we'll move to the key pieces of the Information Marketplace, on which its future will be built.

One more thing before we get started: The book intentionally mixes technical and humanistic views, written in a way that both techies and humies can understand. To do otherwise would be to paint a picture of a world that can no longer be fully understood from either one of these artificial and insular extremes. So if you feel that you are already a hard-core techie or humie, please do not skip what you may fear is "outside" your specialty. It is my hope and expectation that you will happily discover within you your forgotten half—alive, well, and craving to be stimulated.

Finally, throughout the book, many of the examples I draw upon are from my immediate surroundings at MIT. This should not be misinterpreted. Great and pioneering innovations in all of these areas have been made and continue to be made by many good people in many great organizations throughout the world. I try to mention as many of them as I can as we go along, though I could never cover fairly that huge and rich terrain. In explaining the various technologies, I have opted to use the work with which I am intimately familiar, so I can convey to you with the right mixture of accuracy and passion these important endeavors and their consequences. I apologize in advance for this partiality, which in any event does not change either the depth or the breadth of the message.

It is 1964 and I don't have a single strand of gray hair. I am a student at MIT, carrying out computations for my thesis on one of the world's first time-shared computers—a central brain wired to a small number of brainless terminals that allow individuals to share the processing power and memory of the main computer. Of course, I am actually working on a time-shared computer, on the ninth floor of MIT's Tech Square building. And of course, it crashes on the average every fifteen minutes. Such a crash has just happened, and a mad scramble takes place as I and a dozen other students try to catch the computer's attention.

The first crime of the Information Age is about to happen. It will be perpetrated by a frustrated eighteen-year-old and will cause acrimony and, some hope, vindication. But before describing this historic transgression, whose repercussions have yet to be fully played out, let's quickly consider the conditions that led to it.

Pioneered by John McCarthy, now of Stanford University, and Fernando Corbató of MIT in the early 1960s, time-shared computer systems were the opening lines of a fast-moving play that, three decades later, would lead to the Information Marketplace. For the first time, these machines created a community of people who centered their activities around a computer. And even though the community was small, the time-shared days raised the earliest indications of what might happen when millions of people and their computers would form the communities of the twenty-first-century Information Marketplace. Many of the exciting things that a small group of researchers observed in those days are being rediscovered each day, as more and more people turn to the magic of personal computers and computer networks. Some of these early experiences were so rich that they still have not resurfaced in today's setting!

Time sharing was a bright dawn after a fifteen-year-long night during which anyone who wanted to use a computer to solve a problem had to punch holes in each of several hundred paper cards, stack them, feed them into the computer's card reader, and then wait for hours to get the results. Time sharing started as a way to get more value out of big, costly mainframe computers, which cost around $5–10 million in today's dollars. It made economic sense to switch the machine's attention, in round-robin fashion, among some thirty people sitting at their own terminals—first a clunky typewriter the size of a small desk and later an actual terminal with a screen that looked like today's personal computer. The computer was supposed to be so fast that each user would feel that he had the machine's complete attention—the kind provided by a lightning-fast and attentive waiter serving several tables.

In reality, the computer was not so fast, and the people who used it commanded it to do a lot of hard work. As a result, after hitting the Enter key people had to wait half a minute or longer for a response. To be sure, half a minute was considerably shorter than the six-hour waits of the punch-card days. We had traded in our horses for cars. But once you got used to interacting with the computer, you changed your expectations. The delay was like pressing the accelerator pedal and waiting for half a minute before the car moved forward.

Predictably, the pompous, impatient higher-ups in the university research establishments decided that professors and privileged individuals should have the power to bump students and other lowlifes from the computer by merely logging in. If the host computer was maxed out and Professor Wunderbar signed on, some student's terminal would be instantly cut off. Students like me had vivid daydreams of revenge when, without warning, we were suddenly flashed the typewritten message, "You have been preempted by a privileged user," and our keyboard went dead.

It was in that info-feudal setting of haves and have-nots that the vexed eighteen-year-old whom we'll call Ben Bitdiddle quietly invaded the computer's inner sanctum one night, where the names and privileges of users were stored. Having broken in to the password file, Ben, a modern-day Robin Hood, proceeded to simply reverse the privileges, so the lesser people now had the power to bump the higher-ups. I noticed this with disbelief and elation the next morning, as I and my fellow students logged in and ousted the faculty, including the computer lab director. After some gloating, laughing, admiring, and pondering what this new genre of mischief might really mean, the lab administration called the inevitable meetings to decide the young man's fate. After all, the right example had to be set. Eventually, Ben got a mild reprimand—and computer crime was born.

This lesson—*that info-feudalism by a few princes is intolerable to the serfs of a computer community*—has yet to be learned. In 1995 the U.S. Congress proposed tougher censorship rules for online information services, and the services themselves tried to control the actual content of discussions on their electronic community bulletin boards, citing good taste as their reason. These actions met with loud public protest. The companies quietly pulled back. Ironically, when Congress passed legislation that attempted to control the electronic publishing of text and images deemed offensive, one of the services blackened the backgrounds of its subscribers' screens as a protest of its own. In 1996 the same scene was repeated in Australia. The info-barons who are in a position to collect money from their users or their users' activities should stand warned that revolt surely awaits if they try to take undue advantage of their power. Politicians should take note, too.

Not all "crimes" of the time-share era were discovered. For the benefit of this book, Tom Knight, now a distinguished senior researcher in MIT's Artificial Intelligence Laboratory, admitted to me that three decades ago he surreptitiously introduced the capability for a user to get higher privileges by typing the characters *getcom* after his name when logging in. Tom's violation of the system was never discovered. This may seem like a small hack, but it leads to another important lesson: *computer crimes will certainly be attempted in the Information Marketplace, and some of them will never be detected.* We'll talk more about computer crimes and the means to combat them.

From today's perspective, the most significant thing that happened in the time-shared era was the birth of the fundamental force propelling us to the Information Marketplace: a gradual but relentless shift from sharing the high cost of a computer to sharing information. This began as faculty shared with their colleagues papers and computer programs they wrote. It quickly expanded to lists of recommended restaurants and wines and lists about other common interests where people could offer their accumulated personal experience to the community.

The significance of the community that began to form around the time-shared machine was not immediately obvious to us. We were too excited by the new world we were mining, and we were having way too much fun with our community games. In Maze Wars you drove your own vehicle alongside other people driving their vehicles in the same maze, from the comfort of your keyboard and display screen. When you took a turn and saw someone in front of you, it was shoot-out time. A win went to the driver with the fastest reflexes.

In a variant of the word game Perquackey, the host computer would randomly generate ten letters and show them on everyone's screen. We all had three minutes to compose the largest number of English words we could muster. When one user came up with a word after a long pause, a burst of activity would follow as everyone rushed to crib variations.

Then there was Horse Race, devised for the computer faculty's children by my student Steve Ward, who would later become a famous professor. During the week the kids would trade horses for breeding purposes. The inventive programming ensured that when you crossed two horses, the filly or colt would inherit randomly most of the parents' characteristics but would also develop some unpredictable behavior . . . sometimes good, sometimes bad. The children would scramble to rent fast horses from their peers that they could pair with their own best horses to breed really racy offspring. They had to haggle, pay, and collect for such services, and their computer bank accounts would go up and down. On Sunday the kids, sitting in their homes spread across a ten-mile radius, would meet over their terminals to race their best horses. The grand event unfolded painfully slowly on a horseshoe-shaped racetrack shown on everybody's display. But a race it was, with all the adrenaline and screaming you would expect, as the children rooted for their horses to win the purse.

The social dynamics were so realistic that at some point a horse thief emerged. The son of a faculty member, he would surreptitiously visit the computer files of other kids and steal a really fast horse long enough to breed it with his own. He was eventually caught, and his father "grounded" him by not letting him use the computer for a very long time.

As we played these games and engaged in our more serious endeavors, our behavior began to shift: from a bunch of individuals using the time-shared computer to fulfill our own needs, we became a community of people whose activities centered around a computer. It was the origin of what is beginning to happen on a global scale today—the Information Marketplace. I say "beginning" because a great deal lies ahead. First of all, only one half of 1 percent of the world's people are interconnected. And very few of them play games. Even in the small world of games, the interactions that take place on today's computer networks are strikingly lifeless by comparison. Most users have yet to experience the rush of a truly interactive mass game. The games, in turn, are patiently waiting to be rediscovered in settings that will entertain thousands instead of tens of users—and with color graphics, sound, and high speed.

The few examples I've mentioned cannot possibly reproduce the community ambiance of the time-shared period that was taking place within groups at Berkeley, Carnegie Mellon, Stanford, and MIT. After supper on a typical weeknight, any handful of us would meet on the system, typing from our studies and living rooms. We would exchange notes or some online talk, able to read what everyone else was typing. I and others then might join an ongoing game or work on our own projects. Because most of us were logged in all evening, we could alert someone by beeping their terminal, and they would usually join in whatever activity was in progress. New activities were strong draws. No one viewed these interruptions as violations of privacy; if you were busy you simply said so, and everyone would leave you alone.

The fledgling time-shared era taught us that *people value greatly the ability to form a community bound by the sharing of information and are willing to readily integrate new information-driven activities into their daily lives.* This is perhaps the biggest lesson for the future:

people find the Information Marketplace socially and professionally acceptable, even desirable. That half-percent of the world who are interconnected will surely grow to perhaps 20 percent during the next century.

No More Buses

In the early 1980s personal computers emerged, and you no longer needed to share a machine because of cost. Getting your own machine was like buying a car: you'd never again have to wait for the bus, much less get bumped off onto the street. But because of technical limitations, these independent computers could not easily share information. Once people began driving their own cars, they no longer had people beside them to talk and play with. The gain in computer cost-performance came at the expense of losing our online communities.

Researchers throughout the world began tackling the techniques that would make large-scale sharing possible. But not everyone thought that independent computers could be effectively linked together. IBM, even though benefiting from sales of its "PC," could not believe that the mainframe computer and its terminals could be replaced by "distributed" personal computers with no central authority to control them.

At a dinner meeting held at MIT around 1978, the late Bob Noyce, a founder of Intel—then still a small company trying to integrate the processing units of the nascent personal computers onto a single computer "chip"—asked Bob Evans, head of a major IBM division, "What will you people do when the world is crawling with microprocessors and everyone is building their computers around them?" The charged silence that followed was broken by Evans's tongue-in-cheek response: "I guess we'll fold!" Roars of laughter exploded from everyone in the room. No one envisioned the thousands of people IBM and other makers of large computers would have to lay off years later, having steadfastly refused to recognize the worldwide shift from a few big machines to masses of small machines as a result of the rapid development of the microprocessor by Intel and others.

My colleagues and I had several heated arguments with IBM about the interaction that could be developed among interconnected, independent computers, starting in 1977. The most that they would concede was that the dumb terminals dangling on a "host" mainframe would become increasingly intelligent and perhaps *someday* (spoken with a lack of conviction that made it sound more like *never*) they might become *co-equal,* as they called the relationship, to the central machine. We believed, instead, that starting from full equality we would get better and even different results.

As early as 1976 our Laboratory for Computer Science had set the goal of achieving the same level of sharing and coordination among distributed independent machines that time-shared computers had made possible a decade earlier. Unfortunately, neither we nor anybody else have yet achieved that goal, notwithstanding all rhetoric to the contrary. Today's PCs are networked well enough to navigate the Web, swap documents, and send electronic mail—a useful service but a technical no-brainer that transports uninterpreted symbols from one machine to the next. The difference between a time-shared computer and a network of PCs is similar to that between a company where everything must pass through the CEO and a company where employees can interact directly with one another. The former can be run more tightly than the latter but has no hope of growing beyond the number of employees that a single person can manage.

The personal computer weakened the community bond that had formed in the time-shared era. But it was an essential step to break the notion that a centralized machine was needed to coordinate and control people at distributed terminals. The Information Marketplace is the next step. It will rebuild the notion of community, this time among millions of people at powerful machines. And the historians will say that the world moved from computer autocracy to computer democracy.

The Commercial Siren Song

Even before the PC mind-set began to take root, the economic implications of a networked community started being noticed. In the mid–1970s several efforts were launched to profit from the billions of dollars that would become available if computers and communications could somehow merge. AT&T began using its profits from telephony to get into the computer business. IBM, in immediate defense, decided to do the converse: it formed a partnership with Comsat and Aetna, and together spent more than one billion of today's dollars to launch Satellite Business Systems (SBS)—a joint venture that was to provide a communications service for speech, data, and images among the world's businesses.

Early corporate efforts like SBS were based on the realization that voice and computer data could be represented digitally and shipped over the same pipe—in this case a satellite. More broadly, telephone and computer companies began seeing their worlds as converging, and they naturally saw greener grass in their neighbor's yard.

These efforts did not pan out. On the surface, the AT&T divestiture caused it to get out of the computer business, which, in turn, prompted IBM to exit the communications business. At a deeper level, the reason the early ventures, and similar ones in Europe, failed was because *they jumped into areas where they had no core business experience and aspired to offer everything:* hardware, software, communications, and services. This is another important historical lesson that has yet to be learned, and companies in the 1990s stand to waste a great deal of money pursuing the same folly. Recent deals between big media, cable TV, and telecommunications firms are based on an intent to offer total infrastructure and services to their customers—as absurd as a company declaring that it will produce all the highways, cars, signaling systems, repair services, and roadside cafés for "the complete modern transportation system for your home and family."

In addition to not "sticking to its knitting," a company aspiring to this broader goal is often confusing two distinct roles: generating a market where information is bought, sold, and exchanged versus offering the goods and services that are traded. The two roles differ radically—like building a fish market versus selling fish. *Today's companies will do well to decide on which side of this fence they want to be.*

While manufacturers were attempting to cross disciplines in the late 1980s, "information utilities," also known today as service providers, emerged. CompuServe, Prodigy, and Dow Jones, and later America Online, offered information they thought large numbers of office workers and homeowners would want, could easily access from their phones, and would pay for: stock quotations, analyses of company performance, retrieval of published news articles, travel information, electronic mail, and electronic bulletin boards where people could leave and receive messages from many others. Later, the offerings grew: airline and hotel reservations, some limited shopping services, chat groups where almost real-time conversations could be sustained by typing, even software that could be copied, or "downloaded," onto your computer for later use. These activities were sufficiently useful to justify paying a monthly service fee of ten to twenty dollars.

Having started in the early 1980s, the French national telephone utility, France Telecom, took a different approach with a service called Teletel, which eventually became better known by the name of its terminals—Minitel. By the mid-1990s, 6.5 million terminals had been installed next to telephones in homes, businesses, and public places, subsidized by France Telecom with an initial investment of $2–3 billion. With a single phone number and five keystrokes, a user can reach any one of Minitel's growing list of 23,000 different services (in 1996). The most popular service is the electronic White and Yellow Pages—which makes sense since Minitel was built in the first place to replace the massive phone books. Next in line are all the services for scheduling transport by aircraft, ship, truck, and boat; for ordering tickets for the theater, the train, and all sorts of events; for individual bank account management; and, of course, for *romance,* reminding us that the Information Marketplace is destined to reflect human nature.

These services are invoiced on the purchaser's phone bill if they involve information and on credit cards if they involve purchase of goods. Starting in 1996, smart cards, which we discuss in chapter 4, are being used to authenticate the user's identity so as to avoid the security risk of credit card numbers flowing over the wires. At the same time, France Telecom is trying to expand the text-based Minitel to include video and voice transmission. Though this is undoubtedly an important step, it is by no means the ingredient that will turn this service into a true Information Marketplace. Minitel hasn't quite made it, because the French phone company retains the role of central agent.

The U.S. service providers do the same. An insurance claims adjuster in Virginia Beach cannot provide his services on America Online. A database operator in San Antonio cannot distribute his prized information over CompuServe. These utilities are not equipped to act like a market for your goods and don't at this writing support the technology needed to deliver them. If the service you want to sell is big enough, however, you can sell it to them wholesale or provide it through them. They will then resell to their members, retail, or let them connect to your service. That's not good enough. *The Information Marketplace should make it possible for anyone to buy, sell, and exchange his or her goods without having to register with or be controlled by a central authority.*

Despite Minitel's limitations, it has a head start among the information utilities toward brokering services. And yet, neither it nor any other service provider has the critical mass of subscribers needed to provide global reach. As a result, the utilities started making deals with one another. And they have provided access to the Internet and the Web. Will they end up being the winners in tomorrow's Information Marketplace? I don't think so. Access to the Web is just as easily (maybe more easily and more economically) provided by the telephone and cable companies. They are more likely to end up as the equivalent of large department stores and shopping malls in a larger free market economy, providing lots of different products and services under one roof, especially to people who are not inclined to walk around the market to conduct all their shopping. If that happens, they may level out with a market share comparable to that which traditional department stores have of the total market for goods.

Arpanet, Internet, and the Web

To understand the Arpanet-Internet-Web chain is to understand how we got where we are today and, more important, how we are poised to continue this evolution toward tomorrow's Information Marketplace.

Time sharing was jump-started after the surprise 1957 launch of the Soviet satellite *Sputnik,* when the U.S. Department of Defense (DOD) established the Advanced Research Projects Agency, or ARPA, to strengthen national security through far-reaching research. The DOD viewed the computer, which was still a laboratory curiosity, as potentially important to military command and control. But the director of the agency's Information Processing Techniques Office, the late J. C. R. Licklider, had a broader view. A psychologist, Licklider saw a new era in which computers and people would act in concert. The thought was revolutionary and, for many, preposterous: I remember well Licklider telling us these ideas in an after-dinner speech in 1964. Respected scientists were rolling their eyes and making surreptitious shoveling motions with their hands. This reaction was consistent, and a lesson for all of us when facing new developments; *hardly any major innovation was welcome when it first appeared. Yet after a while, as the philosopher Arthur Schopenhauer notes, everyone would agree that "it was all along an obviously great idea."*

Our very own Laboratory for Computer Science was established with ARPA funding, and reflected this purpose in its name— Project MAC (for Multiple Access Computer). ARPA also selected Stanford University and Carnegie Mellon, along with a few other organizations, to pursue the exciting new prospects of time-sharing computers and making them behave more intelligently. The $10 million or so per year (in today's dollars) for each group was devoted to basic long-range research and became a major force in the evolution of computer technology.

As a student at MIT, I found that my roulette wheel had hit on one of the three universities ARPA would fund. Even more luckily, I joined the faculty in 1964 and ten years later became director of Project MAC (which I renamed the Laboratory for Computer Science to make it sound less like a hamburger and more like a research center). I consider this the luckiest card of my professional life—it gave me and my colleagues the rare privilege of growing up next to the relentlessly exciting computer field, marveling at its capabilities . . . while being under the impression that we, too, had our hands on the steering wheel.

ARPA's contribution to the new world of information was spectacular. We can credit its investment with somewhere between a third and a half of the major innovations in computer science and technology. These include time sharing, computer networks, landmark programming languages like Lisp, operating systems like Multics (which led to Unix), virtual memory, computer security systems, parallel computer systems, distributed computer systems, computers that understand human speech, vision systems, and artificial intelligence, an endeavor responsible for understanding and emulating human intelligence by machine.

The commercial world contributed the microprocessor chip (Intel), the shift of computers from laboratory devices to retail appliances (Digital Equipment Corporation), and the personal computer (Xerox, Apple, and IBM—in that order!). Industry (primarily through Microsoft) also developed the shrink-wrapped software we all take for granted. Today, the economic benefits of all these innovations account for 10 percent of the world's industrial economies, nearly $2 trillion a year worldwide. Not a bad return—100,000 percent—on the $1 billion (in today's dollars) that ARPA spent on computer research during its early years to fund "half" of these innovations!

Some of the "histories" floating around suggest that the next big step—the creation of the Arpanet—was done by the military to decrease the vulnerability of concentrated information sites. That reads well, but it is only part of the story. The success of time sharing created mounting financial pressures on ARPA, as every research group funded by the agency asked for more and bigger expensive computers of their own. ARPA sought to leverage its funds by having the groups share distant machines. At the same time, technical factors were beginning to point to the exciting prospect of connecting machines together. Based on some blend of all these reasons, Robert Taylor and Larry Roberts, who later on held Licklider's position, stimulated and encouraged research and projects in computer networking. I remember thinking it was a crazy idea. But I and my colleagues from other universities were used to crazy ideas becoming the engines of important innovations, so we tried to make a winning play out of this apparent nonsense. The result was the Arpanet, the granddaddy of today's Internet.

The prototype network was built in 1969 by Bolt Beranek & Newman. After some tuning and ironing out of (many) bugs, and a public demonstration in 1972, the demand for the Arpanet started growing. By the mid-1970s a few military sites, plus about twenty universities, were linked to Arpanet, and demand started growing. The network's reach now allowed us to log in to distant machines. But in a familiar replay of what happened with the cost-sharing motive that led to time-shared computers, the objective of utilizing remote machines that had led to the Arpanet didn't pan out. The absence of shared communications standards made efforts to explore and use the rich worlds of other machines exercises in guessing arcane acronyms and programming commands—like trying to navigate uncharted terrain with signs written in unreadable symbols. Fortunately, the unplanned consequences of innovation once again ensured that progress would be made—we were soon exchanging electronic mail and transferring technical papers and programs freely and rapidly with colleagues thousands of miles away.

I hope the people and organizations responsible for funding and leading research in computer science and technology throughout the world will learn these powerful lessons from the time-shared and networking eras: *Don't get too fixated on narrow objectives. Be flexible and have faith in good people working on new problems. They will come up with great results that neither you nor anyone else anticipated.*

Another vital lesson for today and tomorrow is to ensure that our computers avoid the difficulty Arpanet users faced in dealing with different and mutually incomprehensible systems. Major players of the 1990s who keep extolling fiber optics, real-time video, virtual reality, multimedia, and electronic commerce will discover that *none of this awesome stuff will be useful unless computers and software at diverse sites can "understand" one another,* at least at a rudimentary level, so they can carry out the desired transactions among them. Without explicit solutions to this problem, our high-tech computer systems will be like a roomful of people addressing each other in melodiously accented languages nobody in the room can comprehend. Hardly anyone today is paying attention to shared conventions that will allow interconnected machines to understand and work with one another without the constant intervention of a human being.

Besides reestablishing electronic mail and establishing file transfer among different computers, the Arpanet created a new breed of people. We called them tourists because they were allowed to log in to our machines from far away and to use our advanced computing resources free of charge when our researchers were not using them—typically between 1:00 and 5:00 A.M. They played games, programmed, and communicated with one another and with us. At one point we had 1,500 tourists registered on our four large computers. Stanford and Carnegie Mellon had similar numbers. Our "security system" was simple and effective: to become a tourist, you had to know someone at one of the university research centers who would vouch for you. Several tourists, then in their teens, are now celebrated information-technology leaders in academia and industry. Another good lesson to be heeded by those who feel that today's youth are mindlessly surfing the Web.

As director of a major research laboratory, I was constantly in fear of tourist behavior that would get us in trouble. A frightening scenario, which happened more than once, went something like this: A few kids would pool information on different wines and vintages for a wine list. Some congressman who was not overfriendly toward ARPA's budget would get wind of this activity and call on the carpet the director of the agency, complain that precious tax dollars were being used for frivolous purposes, and tell him that in the next budget cycle ARPA's funds would have to be cut by maybe $50 million. The director, in turn, would call me and threaten that unless this nonsense stopped immediately a few million dollars might be trimmed from the laboratory's funding. I would counter politely that if our researchers were shackled they would not produce the great results expected by the Department of Defense.

None of these potential threats and counterthreats resulted in anything serious. People in all camps understood the value of leading the evolution of the computer field and of developing young people's expert knowledge of it. After everybody made the noises expected of their positions, life went on.

Regretfully, that wisdom is now in jeopardy in the United States. In the early 1990s, a turf battle over control of budgets caused a proposed billion-dollar cut on the research side of the DOD budget. The cut was eventually reduced, but it is telling of a pervasive trend in a new era in which government officials have become intoxicated with accountability and consequently with the micromanagement of the nation's research enterprise. The people who appropriate funds to government agencies try to score points with voters and the press by cutting expenses that offer no quick or easily defined payoff. At its extreme, this fashionable yet uninformed slashing is not far from trying to save electricity by shutting down a hospital's life-support systems. These well-meaning people talk about the need for ARPA and universities to "trim down" and become as "productive" as the nation's companies. Clearly, they are oblivious to the 100,000 percent return on investment that ARPA made and continues to make possible. *They should study it well, along with the ways universities have helped the United States achieve and retain its primacy in one of the few areas in which the nation still excels.* Let's hope the hula-hoop syndrome, which ensures that no fad lasts more than a few years, will cause this activity to pass. Better yet, let's hope for wisdom!

Arpanet would lead to the Internet and the Web, setting a course for tomorrow's Information Marketplace. Soon after its inception, the Arpanet became busier than expected, so the community started talking in earnest about networking different networks. And ARPA legitimized these aspirations by establishing an internetting research project to that end.

Three key activities "led" from the Arpanet to today's Internet. The first was a 1974 paper by Robert Kahn of ARPA and Vint Cerf of Stanford on what eventually became the TCP-IP protocol—a method of addressing many different networks using a long number. The second was the formation of a chain of grass-roots groups that would steer the Internet standards forward, starting in the early 1970s. This approach marked a major break in the way standards were formed. Instead of the top-down processes that took years to gel, the new groups operated in an informal manner, seeking advice, trying a quick idea here, giving out some code there, to see if it "took" and until it "felt right." This seemingly anarchic process moved the networking effort steadily forward. Interestingly, the Web would follow the same path, suggesting that *we should legitimize this new way of developing standards.*

Cerf headed the first such Internet group until 1982 when he left ARPA to join MCI. David Clark, a senior research scientist at our laboratory, then took over. At the time, the Internet was still small, with tens of networks and hundreds of computers. Enter the third key event: the emergence of local area networks (LANs), which hooked computers and workstations together within a building. The LANs became possible largely because of the invention of the Ethernet, by Bob Metcalfe, an alumnus of our laboratory, and his collaborators. Because of the rapid growth of personal computers and workstations in the 1980s, LANs caught on like wildfire all across the United States and placed a huge demand for connectivity on the burgeoning Internet. Whereas the group under Cerf was concerned with the early evolution of the Arpanet, under Clark it faced all the problems of scaling up to much larger numbers of users.

Outside ARPA the world was also moving toward networking. The military, wanting to have its own high-quality network with good service and reliability, pulled away from Arpanet in 1983 and formed a military copy called Milnet. In the mid-1980s the National Science Foundation, in its quest to enhance the infrastructure for scientific research in the United States, established NSFnet to interconnect the supercomputer sites that it had established and to enable scientists and technologists to work more closely together. The NSFnet propelled the Internet forward with a massive expansion of the backbone that connects key transcontinental sites and with other networking innovations. U.S. agencies like NASA (National Aeronautics and Space Administration) and the Department of Energy joined in, along with organizations and institutions from other countries. In the mid-1990s there were a number of Internet high-capacity service providers, including AT&T, MCI, and Sprint and UUNet and PFINet. They form the *backbone* of the U.S. Internet. By the late-1990s, the baby network that was once destined to have no more than sixty-four *computers* sported over 200,000 interconnected *networks* serving some forty million users—and it is growing at 100 percent per year!

Despite its spectacular growth and usefulness, the Internet did not become a widespread cultural phenomenon until the Web—

and the browsers like Mosaic and Netscape needed to navigate it—hit the streets. The browser was pioneered by a twenty-two-year-old programming athlete, Marc Andreessen, first at the National Center for Supercomputing Applications (NCSA) and then at Netscape Corporation. The Web holds a hidden lesson of its own: Technical proficiency means nothing to the general public. Ease of use and ease of posting their own information is what matters to users.

An individual or organization that wishes to tout its wares on the Web establishes a *home page*, where it can present information. Certain key words and images on the page are highlighted. A user who clicks on any such highlighted object is transported to supporting details, photos, sound bites, data—or to a completely new home page from another user or organization, which could be in West Los Angeles or West Africa. The new page also has highlighted words and images that, if clicked, open more doors to even more chambers of information. A simple click can take you to the White House, the Vatican, a video sex palace, or the trading desk of the Tokyo stock exchange.

Imagine millions of people organizing their information under these simple rules of association. That is what's happening today as people and organizations from all over the world follow their commercial or personal siren song to show off and share their wares. The resultant web of home pages becomes big, complex, and interesting because it is created, organized, and interlinked by all these different people. When Tim Berners-Lee invented the Web, he envisioned it as a growing superhuman "brain" formed by linking together a lot of individuals' knowledge around the world.

The Web was actually invented by combining two approaches: One was an addressing scheme, like a street name and number, for locating files, pictures, audio, and video anywhere on the Internet. The other was a simple language for assembling such information into home pages on any kind of computer and a set of conventions for linking and transporting any such information across the Internet. (See also the appendix.)

Simplified, these key contributions blend two old ideas: networking and hypertext. We've discussed networking. Hypertext is a new kind of electronic book with highlighted words: Click on a word and you are presented with an elaboration—be it text, graphics, audio, or video. If you want to probe further into something you see next, just click on it. And so on. The now-popular CD-ROM encyclopedias are the best examples of hypertext documents. The genius of the Web lies in extending this hypertext notion from a book to the entire community of networked computers. That's the great lesson for the future: *keep the infrastructure simple in concept and easy to use and share, and it will spread rapidly through the world.* Like many great innovations, the Web's simplicity became apparent only in retrospect. It was not foreseen or invented by thousands of techies who knew all about networking and hypertext . . . except for Tim Berners-Lee.

The Web also provides an important negative lesson. Exciting as it is, it is still quite chaotic and still far from a true information infrastructure. The best way to find anything is to explore. This can take hours, like walking through a huge market where there are thousands of streets and alleys full of stuff. Most of it is boring or irrelevant to your needs, and yet you must wade through it all to find the few jewels.

To help you knife through the jungle, systems called *search engines* are being offered. With names like Yahoo and Alta Vista, they crawl all day and night on the Web, investigating every highlighted word they find and storing in a gigantic index the locations where they find it. You type in a key word or phrase you want, and they rapidly list all the home pages they have in their index with words that match your magic words. If in July 1997 you used Alta Vista to search for "Mary, Queen of Scots," you got 400 hits. If you searched for "baseball," you got over 500,000 hits. That's nice, but then you would still have to browse manually through these half million sites to find what interests you.

For a functional Information Marketplace, this chaos has to yield to guides and yellow pages and software that won't just assemble mindless pairings of words and sites but will present you with a tidy, velvet-lined box of jewels—answers that truly and closely match your questions. This development will be difficult and will require human editing because machines are not smart enough to do this kind of organizing by themselves. But it is critical. Much more will have to be done to move the Web toward an information infrastructure too, as we discuss in chapter 4.

In retrospect, the evolution of Arpanet to Internet and the Web provided the connectivity to (almost) re-create the communities of the time-shared era. What was once the province of a few dozen people is now becoming available to tens of millions, who can engage in the useful buying, selling, and free exchange of information with one another. This is the rudimentary machinery on which the Information Marketplace will be built.

War of the Spiders

Now that we know where the Information Marketplace comes from, let's consider the key actors who aspire to shape its future, along with their strategies.

To a media company like Disney in the United States or Bertelsman in Europe that owns cable TV channels and thousands of movies, recordings, newspapers, and magazines, richer content is the fuel that will propel the Information Marketplace to maturity. To a telephone company like AT&T, Deutsche Telekom, or NTT in Japan, the ability to rapidly and inexpensively ship information from anywhere to anywhere is the key ingredient of the Information Marketplace. To a software company like Microsoft, millions of interconnected computers can be put to good collaborative use only if they are fed imaginative and useful software that links them up effectively—"killer applications" or "killer apps," as they are known if they become best-sellers. To a computer company like IBM or NEC in Japan, the Information Marketplace is little more than a billion interconnected computers within a decade.

These companies are attracted by a powerful economic supermagnet. And each of the companies sees itself as the dominant player, needing only a few deals with the other "subordinate" firms to become the driving force behind the Information Marketplace.

Media companies often own the means of distribution—the television cable that reaches many of our homes. They see their job as owning or buying information wholesale from the producers of movies, the news, and perhaps some day the L. L. Bean or Harrods catalogs, repackaging it, and delivering it by predominantly one-way broadcasts to millions of consumers. They don't have experience in carrying other people's information around, and they don't have any desire to spend their money doing so. Who can blame them for that? To be fair, some of these companies do not share that mind-set and have actually begun to plan and install two-way video modems: Continental Cablevision's Highway-1 service aspires to offer a very impressive bidirectional service. But the broadcasting mind-set was still prevalent in the mid-1990s.

This tendency evokes the image of a spider that *controls the flow* of information, amply from the body to the legs but only minimally in the other direction. If these spiders operated the highway system, there would be only one-way streets leading out of their central hub. This would bring asphalt to every home and office, but it would hardly be a highway infrastructure, because it would be nearly impossible to go from one place to another. Unfortunately, the cable that passes through all the homes in your area is the same wire. So if you tried to use a TV channel to sell your own videos, you would be blocking that channel from everyone else. If only a few people went online this way, the system would run out of channels.

Some telephone companies are also spiders, but of a different kind! They are after *control of the services* that will be offered. The government-run telephone administrations in some European and Asian countries see themselves as information utilities that will provide many of the useful services needed by their customers. They want to provide not only the highways but also the cars, gas stations, and restaurants. This would not be a true Information Marketplace, because many of the services that would be bought and sold would be theirs, not ours. To be fair, other telephone companies see their role as they should—providing an infrastructure that transports information. Even among the more enlightened quests, however, we can still discern the spider syndrome—a desire to control what goes on by favoring sales of their own new services.

The software and computer companies form another group of spiders, which will try to control the platforms close to the user. They hope to dominate the browser and desktop software—the doors and windows to the Information Marketplace. And they will offer new hardware and software products that take advantage of the distributed computer environment. To protect their market share, many of them will be tempted to limit your ability to access competitors' products and services that are not compatible with their own products. It would be like providing a road system controlled by a single automaker that would let you drive the company's different cars but make it hard to drive other manufacturers' models.

Eventually, the warring spiders will have no choice but to surrender their spider tendencies and to adopt the Information Marketplace. I have great conviction in this statement, because every large and small organization and every individual human being is a potential supplier and consumer of information and information services. We hundreds of millions of people and organizations will not stand on the sidelines, muted by a few companies trying to corner the Information Marketplace. We will favor with our money those companies that give us our voices and help us work, live, and play in the same arena. And the Information Marketplace will not be owned by any single entity, but by all of us!

This marks a major difference between the Information Age and the Industrial Age. We can't make automobiles or produce goods at our homes. But we can certainly create information and sell our office work. That is the bulk of what will flow in tomorrow's Information Marketplace. Before such an avalanche of market pressures, the spiders will have no alternative but to help build the information infrastructure if they want to stay in business. Engaging in wars, in the meantime, will cost us and them a delay of perhaps a decade, and they will waste piles of money.

Among the spider wars the most important ones will be for control of the pipes that transport our information down the street or across the oceans. That's because many of the useful things that we will want to do in the Information Marketplace will require a communication infrastructure that can carry information rapidly, reliably, inexpensively and with the widest possible coverage of homes and offices. To understand where we are headed and what is likely to happen, let's take a closer look at the strategies and the capabilities of the pipe companies—the telephone, TV cable, satellite, and wireless-system carriers of information.

Battle of the Pipes

Many forecasters claim there will be one technological winner. Some say it will be fiber optics. Some say coaxial cable. Others predict we will all have an eighteen-inch satellite dish on the roof. Some maintain that a specially programmed hardware unit at the end of the old copper telephone wires that enter our homes will be enough to connect us with the Information Marketplace.

Telephony offers the most extensive network of pipes worldwide and is relatively inexpensive and convenient. Unfortunately, it is limited in the amount of information it can move each second—its *bandwidth*. It's okay for transmitting text, but slow for high-fidelity sound, excessively slow for images, and unusably slow for video.

Fortunately, the wires that have been laid at great expense from the local phone exchanges to each home are not so limited. And the long-distance lines between exchanges are already 100 percent glass fiber, which means that they have a very high bandwidth. So the phone companies *could* upgrade the computerized switches in their exchanges, for a few hundred dollars per user, to provide bandwidth sufficient for video sent with newly developed transmission techniques. We consumers would have to buy for our end, also for a few hundred dollars, a smart box with several jacks for different devices, much like today's stereo receivers have jacks for CD players, VCRs, cassette tape units, and speakers. We would then plug in our phones, TVs, hi-fi sets, computers, and specialized devices like personalized newspaper printers. But we wouldn't be able to send or receive at any given time much more than a single video stream. Then, during the next two decades, the phone companies would gradually replace the copper wires, first to the curb and then to our homes and offices, with glass fibers, giving us huge bandwidth.

The world's phone companies have a corner on connectivity. They already switch conversations among some 700 million homes and offices. They are also quite reliable, maintaining service even when the power is interrupted. Your computer is sure to have crashed in the last few tens of hours. When is the last time your phone crashed? Years ago, if at all. The phone companies can readily add an electronic mailbox for each telephone number to receive and store voice messages, images, data, and more. And for people who do not want to pay for or bother with computers, the phone company could offer them basic access to the Information Marketplace for a rental fee or a flat monthly charge.

The phone system can also support the Information Marketplace through cellular communications. Mobile communications have expanded explosively from a few privileged emergency vehicles to almost anyone. With cellular phones, developing nations can become interconnected within their boundaries and to the rest of the world without burying a single wire in the ground. Cellular systems can be installed quickly and at low cost. Cellular telephony has small bandwidth, however, precluding video, at least with current technologies. And it's too expensive to leave turned on for a long time. It is also a bit unreliable for data transmission. Snippets of the communications stream are lost as a moving phone is handed off from one local transmitter/receiver to the next. This is tolerable for a conversation but corrupts data, requiring repeated transmissions. However, improvements are coming well within a decade that will overcome most of these problems, extending the Information Marketplace to less-developed countries and the world's urban regions, albeit still at comparatively slower speeds.

Coaxial (TV) cable is unevenly distributed throughout the world. In 1997, 90 percent of U.S. households could be reached by cable, with 65 percent of them having active cable service. The average reach across Europe in the same year was 27 percent, with nearly full coverage in Belgium, Holland, and Bavaria and none at all in Greece. In Asia and Africa cable TV has barely begun.

At first glance, cable would seem to have a much higher bandwidth than the telephone line, because it can move video. The catch is that the same cable that reaches your home passes through every other home in your area, whereas each home has its own dedicated pair of phone wires that connect to the phone exchange. Digging up the streets to string individual cables or fiber lines to every house is prohibitively expensive. As a result, cable companies are developing ways to squeeze more channels over the same coaxial cable, (through a method called compression that we discuss in the appendix). This expense can be justified because it will allow them to offer movies-on-demand, grabbing a major portion of the $30 billion video rental world market. The investment is only 10–20 percent of what it would otherwise take to dig up the streets. Bandwidth for bandwidth, the investment needed to upgrade and share the cable is comparable to that needed to upgrade the telephone system, with a slight edge in favor of cable in the short term. In the longer term, neither approach seems to have a substantial bandwidth-cost advantage for the Information Marketplace. The advantage lies elsewhere.

The media and cable companies' strength is their ability to provide access to content—thousands of movies, recordings, books, and more. The telephone companies' natural advantage is connectivity and switching. If all the information content of the Information Marketplace were supplied by the media and cable companies, then the two sides—telephone and cable/media—would be balanced adversaries, one with connectivity, the other with content. But in reality, the "content" will be supplied by millions of people and organizations as they buy, sell, and freely exchange information and information work.

The traditional telephony services of the phone companies are already beginning to be challenged by "Internet phones"—software tools that carry telephone conversations over the Internet, with the result that you may be able to talk to anyone in the world for the monthly flat fee of your service provider. With this approach you may also experience new and exciting "telephony" services that combine voice with visual menus and other Internet applications! Though this may work for a few phone conversations, there is no telling what would happen to rates and quality of service if it were to grow. Today's Internet is as unable to support worldwide telephone service as today's telephone system is unable to support a worldwide Internet with hundreds of millions of people staying connected for hours!

On balance, it would seem that by virtue of their experience in providing connectivity among millions of people, phone companies have an inherent advantage over cable companies. Whether they use it is a different story. Cable companies are busily at work, moving fast to address these emerging issues, and in cooperation with long-distance carriers may end up moving ahead of the phone companies.

Another important pipe technology is satellite communications. One geo-stationary satellite—so named because its orbit keeps it fixed above one point on the earth's surface—can support two-way information flows through little dishes mounted outside our houses and offices, all located within the satellite's "footprint," which covers millions of square miles. AT&T is planning to launch its Voice Span system, consisting of sixteen such birds, aspiring to serve tens of million people with hi-fi voice, video, e-mail, fax, and other applications for tomorrow's PC-phones. The Hughes Space Way is a comparable twenty-bird system with similar aspirations. Both systems should be launched by the turn of the century.

Because they are so far away, geo-stationary satellites introduce a quarter-second delay between the two earth stations they connect, which means half a second for a complete round-trip transaction. This is okay for voice and some computer applications, but it is less tolerable by others that need to ping-pong data back and forth.

An interesting innovation in the planning stage that substantially reduces this delay is the Low-Earth-Orbit (LEO) Satellite system; hundreds of satellites would be lofted in lower, fast-moving orbits to blanket the earth. Microsoft's Bill Gates and McCaw Communications founder Craig McCaw, with others, are providing private capital to turn this dream into reality through their company Teledesic. Motorola's Iridium system and Loral Aerospace's Global Star system are other LEO satellite systems with similar aspirations but different numbers of satellites and different frequencies used. Unlike the geo-stationary birds, the LEO satellites whip from horizon to horizon in a few minutes, so they must hand off signals among themselves much as the cellular systems cope with moving users. The concept looks good on paper but will have to be proved in reality before we can see its future niche among the pipes.

The big advantage of satellites is that they can reach anyone in the most remote part of the world, whether on a mountaintop in Nepal or in the lagoon of a tiny Pacific island. There is simply no other good way to do that. They may also be used to service sparse suburbs and to supply occasional high-speed "bursts" of information to people who would otherwise need a low average bandwidth.

While people argue about whether telephone or cable or satellite companies will win the pipe wars, a surprise contender may emerge—the micro-cell wireless communication system. It is a much more densely configured cellular system, in which the cell—the basic unit of area covered by a single transmitter/receiver—is perhaps one city block rather than a few square kilometers. Little transmitter/receiver boxes mounted on every street corner would reach every office, home, and car, as well as every pedestrian, cyclist, and skater in that block. The bandwidth of each connection would be large enough for video because there would be fewer conversations within a cell. And the power required to transmit would be small because the signal would have to travel only a few meters rather than a few kilometers. Most important, no digging would be necessary, so the cost would be low.

The micro-cell wireless entrepreneurs must find a place to mount all the little boxes so they can communicate with one another and their central exchanges. This means that phone and cable companies, which own utility poles and underground ducts, might play a big role in the surprise. It also means that unexpected players like the power, gas, water, and sewer companies, which also own extensive rights-of-way, could take part. Micro-cell wireless systems will have to prove their reliability and efficiency; so far they only work on paper and in small prototypes developed by several startups. Maintenance of the boxes could prove expensive. These systems, too, will require the switching capabilities of the large telephone companies to reach distant sites.

The wars between the owners of the pipes will be hard fought and traumatic. The media, software, and computer companies are used to competing for customers. Their customers, too, are used to choosing among different suppliers for their goods. The telecommunications world, on the other hand, has been a legal or virtual monopoly for decades. Neither suppliers nor customers are used to a competitive environment. This monopoly came about because when the phone systems were being erected, governments had to offer the phone companies guaranteed streams of monthly revenues so that they would supply the massive capital to string up the poles, dig up the streets, and guarantee universal access to every site, however rural.

In the United States, one of the most important consequences of the 1996 Telecommunications Act is that almost anyone will be able to compete for access to your home. In practice only big players like phone and cable companies will be able to afford the high cost of doing so. Privatization and deregulation have also been mandated by the European Commission and by individual governments there. Japan will most likely follow the U.S. example and chop NTT, now the world's largest phone company, into several independent and deregulated pieces. The global privatization and deregulation of communications is inevitable as the world's economies compete with one another, as people seek higher-quality services, and as all kinds of transnational carriers rush to exploit the emerging Information Marketplace.

A note of caution is in order, however: Although it would seem that the era of national telecommunication monopolies will be finished, there is a strong possibility that such monopolies will persist. Deregulation will come and they will vanish legally, but they will reappear as a handful of powerful alliances spanning the globe. That "dance" has already begun.

Exactly how the pipe wars will be resolved is unknown, of course. What we can forecast with some confidence is that the different technologies will contribute to the Information Marketplace, because each has a unique advantage. Don't look for a single big winner, as the hype suggests, but a balance among several winners offering complementary value. We can also anticipate with confidence that bandwidth and reliability will increase steadily during the next two decades because of pressing and growing user needs, expected high economic gains, and the availability of the technology needed to fulfill these expectations.

As users, it is also predictable that we will face a situation similar to today's long-distance telephone wars in the United States. Providers will bombard us with cleverly packaged services at cleverly packaged prices, and somehow we will muddle through the bewildering choices toward a decision. Prepare yourself. No doubt, you will hear from the phone company, the cable company, the satellite company, the wireless company, and new companies that agree to blend all of these pipe services into a single box they will sell you that connects to both the phone line and the TV cable and has a small dish and an antenna on the side! Deregulation will also exacerbate the trend toward multiple offerings and wars among the pipes.

That thought brings us back to the bigger war between the spiders, including the pipes. The terrain is so large and diverse that no single organization or even group of organizations can win it all. Each type of company should focus on what it does best—and do it. And all the companies should work together with their competitors to build (or at least not to block) a shared information infrastructure. This is not an idealistic wish, but a call for practical actions these companies can and should take: cheaper (even free) shared software, widespread compatibility between products and services, and participation in standard-setting efforts among information providers and consumers. *Oddly enough, cooperating on a shared infrastructure is the best these companies can do for a healthy competitive future.* It's not unlike what opposing football teams without a playing field must do before they can compete. With the information infrastructure in place, all companies can then use it to exercise their ample creativity and hard-earned expertise, and even their spider instincts for their own core strengths. There is enough complexity, money, and promise in each of these major areas to keep everyone busy, wealthy, and overflowing with challenges.

The Five Pillars of the Information Age

I have already made a number of assertions about the future and will make many more. Some are derived from today's forefront research, others from the lessons in the history of computing. Many of them are also based on the fundamentals that govern how computers represent, manipulate, and transmit information. Imagine trying to understand the potential of the automobile a century ago without knowing about internal combustion engines, fuels, and tires. You might have been led to believe that some day you would be able to drive from New York to San Francisco in one hour. You might then forecast all sorts of interesting things about the social consequences of the automobile; for example, everyone would be able to commute to work no matter where they lived, so the family would stay together! It's no different with the Information Marketplace.

With this in mind, let me present here a very short summary of the basics so that you will be able to examine on your own what is going on around you today, draw your own conclusions about the future, and judge for yourself the predictions made by me and others. And let me note that this will include a few surprises, for example, the notion that information is not just text, images, and videos but also the active processes that transform these things. Discussing the humie and techie issues together is important. After all, this is a book about the combined forces of humanity and technology. We can't brush aside either one.

The basis of the Information Marketplace is, of course, information. But just what is information?

- The time of day is information—as is tomorrow's weather, a ship's course, and a baby's weight.
- The contents of a typed memo is information—as are the contents of every book written and of the Louvre and every other museum.
- Bird sounds and presidential speeches are information—as are radio shows and all the music ever played and to be played.
- All 20,000 commercial videos and movies are also information.
- The process of designing a house or a car is information—as is most of the office work carried out by hundreds of millions of people.
- Military orders, medical test results, and assembly instructions are information—as are all business procedures and all computer software.
- A computer is described by information—and some day if we crack the biological mystery of our existence, we might be too.

Information is not easy to define, so examples are necessary. What matters to the Information Marketplace, however, are a few key points about the nature of information.

First, humans deal with information on three levels. We receive it with all of our senses. We process it with our nervous system and in a miraculous and largely unknown way with our brain. We also generate it as our brain commands our muscles to speak, scream, gesture, and type.

Second, information can be a *noun* or a *verb*. Text, sounds, images, and videos are information nouns with names like the Bible, the *Marseillaise,* and *Star Trek.* Computer programs that transform text and images and perform work are information verbs with names like Word, Photoshop, and Lotus 1-2-3. Humans produce information as both noun (speech, writing, gestures) and verb (processing of office work using their brains). Whether it is a computer program or a person that takes in information, transforms it, and spits it out, what is being done is *information work.* I will use this term when dealing with information verbs to remind us of the distinction.

As we said in the introduction, most people do not think of information work when they think of information. Yet information work is central to the Information Marketplace and will flow through it like the more familiar passive forms of information; it is a huge part of what will be bought, sold, and freely exchanged. Information work is what Dr. Kane did when he diagnosed the illness of the man at Ruby Creek, and it is what the automated Mercedes simulator did for the man in the Pader-born World Shop.

Third, information is not the same as the physical thing that carries it. A twenty-volume set of books and a CD-ROM can carry the same encyclopedic information. A Renoir nude might sell at auction for $23 million, although the information on the canvas, when printed on a poster, might sell at the museum store for $10.

Once we accept that a huge number of physical things, events, and actions can be described by information, we are only five steps away from understanding the true underpinnings of the Information Age—its five pillars:

1. Numbers are used to represent all information.
2. These numbers are expressed with 1s and 0s.
3. Computers transform information by doing arithmetic on these numbers.
4. Communications systems move information around by moving these numbers.
5. Computers and communications systems combine to form computer networks. Computer networks are the basis of tomorrow's information infrastructures, which in turn are the basis of the Information Marketplace.

If you want to learn more about these things, so that you may better understand the engines, fuels, and wheels of the Information Age, then take a break and visit the appendix, where I review them quickly and, to the best of my ability, painlessly.

Let's now turn our attention to the ways you will interact with this fascinating new technology.

3
Where Person
Meets Machine

Audio, Video, Bodyo

Imagine that since the day you were born, the only way you could talk to your parents, your relatives, and your friends was through a keyboard and a mouse. Not a pleasant thought. Yet we are quite content to restrict ourselves to this unnatural form of communication when it comes to our computers. How come? Presumably, because we believe that this is the best technologists can do given the limitations of machines. Not true! The technology that can let you hold a spoken dialog with your computer in a narrow subject area is already working well in research settings. It also promises to become much more widely used in the emerging Information Marketplace during the next five to seven years, as a result of technical and economic trends.

It's tempting to discuss the human-machine boundary, or interface, as a bunch of gadgets and techniques, where the prime questions are whether the human should use keystrokes, mouse clicks, handwriting, speech, or fancy goggles and how text, icons, and colors should appear on the screen to please human users. Such questions are relevant, but placing too much attention on them is like worrying about whether people should use deeper voices or more body language when they communicate with one another. The overwhelming imperative is to communicate as effectively and naturally as possible, whether writing, speaking, rolling one's eyes, gesturing, or squeezing another person's hand. It's exactly the same with computer interfaces. And as we'll see, it involves the very difficult business of conveying shared concepts.

Interfaces are important because that's where we come into contact with the machinery of the Information Marketplace or, more philosophically, because that is where humanity meets technology. The Information Marketplace will not reach its full potential until the interaction between humans and machines becomes more effective than it is today and closer to human-to-human communication. That's why we discuss interfaces here before we get to the tools we will use in the Information Marketplace.

Besides keyboards and mice, today's interface devices include trackballs and joysticks that move an object on a computer screen, hand-held styluses for handwriting and drawing, microphones that pick up speech, and both still and video cameras for images. However, there are many other devices being developed around the world. They include gloves that let the computer know the precise movements of your fingers. They also include glasses and head-tracking helmets with mechanical, electromagnetic, and optical gadgets that track your eye and head movements so that the computer knows where you are looking. Complete bodysuits that convey the motions of your torso and limbs are not readily available but have been built (in clumsy forms) and will no doubt appear in the future. These same devices will feed information back to you, flooding your senses with spoken information, three-dimensional video, audio, and "bodyo"—tactile impressions that will range from the tickle of a cat's whiskers to being driven into the back of your chair—at least, when the research is done and they become commercial products.

These interfaces may soon allow you to work simultaneously with colleagues around the globe, order food from a French waiter in French even though you don't know the language, and take a dance lesson in your home from an instructor across town. They are also likely to alter your interactions with people on the street and in your home in dramatic ways, which we'll see a little later in this chapter.

Not until we stop marveling at such interactions, and use them as needed to communicate with our machines, will the Information Marketplace become an integral part of our everyday lives. Let's examine the different interfaces to see which ones can really help us, which are likely to be successfully developed, and which may be relegated to their proper place as fascinating but passing fads.

Talking to Your Computer

Though fancy interfaces like bodysuits may provide compelling sensations, all we need to carry out most tasks we will perform in the Information Marketplace is the ability to speak with the computer. Speech is a big part of the interfaces we'll end up with, for two important reasons: It is natural—the vast majority of the time we communicate with one another simply by speaking. And speech is the interface technology most ready to explode for practical applications.

A system that understands speech could dramatically expand technology's role in our daily lives. A navigational-aid program in your car could help you find your way through a city as you drive. Another one in your home computer could guide you through a maze of potentially useful services in the Information Marketplace. Speech systems could act as travel agents, helping you book flights or make car and hotel reservations. They may also be used in athletic event kiosks to answer your queries on the day's latest results, past athletic records, and reporters' commentaries. Then again, they may be used as shopper's aids throughout the Information Marketplace, helping you find out what is for sale or how a product works. They may help you fill out a variety of forms and send electronic or voice mail. Of course, speech systems cannot be used everywhere. In quiet environments or where writing or gestures are more appropriate, other interfaces will be better. Still, speech is perhaps the most promising interface ahead.

A translating telephone is also within the reach of speech-understanding technology. This was the dream of the late Dr. Koji Kobayashi, former chairman of Japanese electronic giant NEC, one of the great engineers of our time. One approach toward this dream has been developed by Dr. Victor Zue, a pioneer of speech understanding research at the MIT Laboratory for Computer Science, and his colleagues. It works like this: Say you wanted to call from the United States an associate in Japan. After you connected with your party, you would speak into your phone in English, and you would immediately hear a computer-generated paraphrase of what you said, to ensure that the computer understood you. At the same time, the machine would translate and present your sentence in Japanese to the other party. If the computer did not understand, you would hear the incorrect paraphrase, hit an abort button, and try to convey your message with a different sentence. As we'll soon see, the discussion must be confined to a narrow domain like setting up meetings. Yet the approach can readily be extended to a conference call, thereby lowering the linguistic barriers to certain kinds of cooperative work. Almost every prototype speech system at our laboratory uses this kind of translation to handle several different languages. You may ask your question in English and get the answer in English, Japanese, Mandarin Chinese, French, or Spanish —your choice.

Another important application, pioneered by Zue and his colleagues, is the Literacy Tutor, a prototype program that hears you read aloud from a book that it knows. It then prompts you how to pronounce words you mispronounced or were unable to read. We often joke that the learners will learn how to read, all right, but may end up speaking with the metallic tones of their automated teacher! People do well with such programs because they do not have to feel embarrassed by their illiteracy in front of another person.

Speech-understanding systems could well dominate tomorrow's interfaces. However, they are difficult to engineer, and the research successes to date are not appreciated. For one thing it's easy to "fake" speech systems that understand only a handful of disconnected words or canned sentences, misleading people into believing that they can handle ordinary speech. For another, there is a legacy of imaginative movies like *2001: A Space Odyssey*, which had speech understanding as a done deal in 1968. Movies like *2001* have jaded us with premature expectations, as I found out in a bizarre incident in 1990. One afternoon a delivery man came into our lab and asked one of the researchers for the location of Cambridge City Hospital. That happened to be one of the questions we often asked our speech-understanding system. The researcher, smelling a sudden opportunity for a live test, dragged the unsuspecting delivery man over to the machine and repeated the query. The man heard his answer, said a casual thank you to the machine, and walked out!

The delivery man had taken the system for granted. He had no idea of the twenty years of research and $20 million it had taken our team to attain that remarkable level of performance.

Because speech understanding by machine is so important to the Information Marketplace, and so that you *will* appreciate its capabilities and limitations, let's briefly explore how this marvelous technology works. Among the many competing technical approaches, we'll focus on one—that of Victor Zue's and his colleagues—which, besides being familiar to me, is considered in the top tier of these efforts. As we do so, we should keep in mind that outstanding research in speech recognition and understanding is being pursued by many scientists, including Raj Reddy and his colleagues at Carnegie Mellon University, researchers at IBM formerly led by Dr. Fred Jelinek, now of Johns Hopkins University, and Dr. Larry Rabiner and his colleagues at AT&T.

For two decades computers have been notoriously poor at comprehending ordinary human speech, in spite of repeated predictions to the contrary. This has led many people to write off the possibility of genuine conversation between people and machines. As a result we hear very little in the press these days about such prospects. As always, however, scientists and engineers march on regardless of what is popular or unpopular, and current speech research has reached a surprisingly good level in the laboratory. Still, speech systems cannot be used indiscriminately in any situation. Though you will be able to use them to book a flight, you won't be able to discuss politics with them, because of the incredible richness of concepts that such a broad topic would entail.

Speech systems will succeed in settings where the domain of discussion is narrow and well structured. They will do better when the speech is a dialog—in the form of statements and responses— not a long singular discourse like dictation of a memo. Programs like Zue's Pegasus airline reservation system can provide results almost as good as those obtained by talking to a human reservation agent—as long as you stick to the topic. Pegasus can interpret some 2,500 words that can be spoken in connected sentences; you don't have to pause strangely between words, as some speech systems require, in order to detect word boundaries. It also tolerates many different speaker accents and requires no speaker training; you don't have to pronounce key words over and over before it is able to understand your voice.

Pegasus brings together several technologies. First, the speaker's voice is sampled 16,000 times a second and converted to a stream of 1s and 0s. Next, the numbers are converted by the computer into related sequences of numbers called power spectra, which represent every pitch uttered, along with the energy and duration of each pitch fragment. These numbers are grouped further into patterns that represent possible phonemes. Phonemes are all the elemental sounds of our language, like *ah, s,* and *d.* Vowels last longer than consonants. Men have deeper voices than women and children. And everybody coughs, sighs, sputters, and hesitates when they speak. As a result, no two people emit the same pattern of numbers even when they try to utter the same phoneme in the same way. To cope with this variability, another set of some forty numbers, called features, are generated for each hypothesized phoneme, which scientists have found characterize the possible phonemes that might have been spoken.

The system doesn't try to guess what the phonemes are yet. Instead, it identifies and stores for future use all the plausible phoneme combinations that might have been uttered. As you speak, the system detects and carries all of these possibilities forward, waiting for more clues to what you might have said.

Next, the system mounts a huge search to compare each of the possible phoneme combinations against its own dictionary, which has multiple entries for each word. For example, Harvard is listed as both Harvard and Hahvahd. This exhausting search takes up most of the time expended by the program, because it compares every one of thousands of possible phoneme combinations to every one of thousands of vocabulary entries.

You think we're done? Not yet!

The vocabulary search eliminates unlikely phoneme combinations. The result is now a dozen or so sentences that might have been spoken. These are examined next for their grammatical and linguistic construction. No one but the most erudite Oxford scholar speaks grammatically in everyday life. Most of us say things like, "... Aaahhh, can you please tell me what [cough] ... ahh when the first flight to Paris is?"

Ungrammatical as this statement may be, it nevertheless has a basic grammatical and linguistic structure. No one is likely to say, for example, "Next when flight aahhh please Paris."

The system filters out grammatically improbable sentences, ending up with perhaps three or four candidates that might have been spoken. These are then presented to the part of the system that understands about meaning in general and, most important, meaning in the narrow domain of flight reservations.

It is here that the system stores things it should remember, along with context information, which will serve as references to other statements. For example, if you say that you are at Times Square and later on you ask where the nearest airport is, the system should still be aware of your location and should respond accordingly. This is also where the system selects the final most-likely-to-have-been-spoken sentence, based on the domain of discourse. For example, depending on whether the domain is ecology or computers, a different final choice will be made between the two candidates "wreck a nice beach" and "recognize speech," which to the computer sound almost exactly the same—as they do to us if spoken in isolation.

Here too is where prosodics come into play—the parts of the sentence the speaker emphasizes. "I am going to Paris," with emphasis on the first word, means that I, rather than anyone else, am going to Paris (and therefore only one ticket is needed, and it should be issued in the speaker's name).

This simplified explanation of how Pegasus works is complete except for the mystical "ear." After many years of studying the way the human ear works, scientists have learned to construct a box that electronically processes sounds much as a human ear does, with its outer flap, drum, those funny little bones, and the cochlea, that spiral cavity full of receptor neurons in the inner ear.

If we place this artificial ear between you and Pegasus, strange things happen: If you are speaking in a quiet room, then there is no difference in overall system performance with or without the ear. If you are speaking in a noisy environment where a lot of other people are speaking, however, then the system performs significantly better with the ear in place. No one technically understands why this should be so—a reminder of our limitations as scientists and an opportunity to marvel at nature's wisdom.

Pegasus is finally ready to decide that what you said was, "Can you please tell me when the first flight to Paris is?" I say "finally," but Pegasus actually performs all the exhaustive steps we just described in about as much time as it takes to say the sentence. It flashes the sentence on the screen for you to see, so you know it understood you as it sends your query to the Sabre airline reservation system (widely used by airlines and travel agents), suitably restated to be understood by that system. Ironically, the machine "fakes" a human as it types to Sabre commands devised for use by human airline agents. When Sabre responds, Pegasus flashes the answer on your screen and also speaks to you through a program that converts text to speech (a much easier task than the other way around). Once you are satisfied with your reservations, you can actually book the tickets on your credit card—a perpetual source of hacks as people in our lab pretend to charge their trans-Atlantic trips to Victor Zue's credit card and then cancel the reservation.

Much of why Pegasus can work so well is the narrow domain of knowledge in which it operates. Keeping the subject limited shrinks the possibilities of what might have been said so that the system can make a confident decision. Consider how many questions you might ask of an airline agent—probably no more than a few dozen generic ones about time, cost, availability, in-flight meals, and so on. You may pose each question in thousands of different ways using different words and sentence structures, but both your statements and your listener's attention will be focused on the small number of targets that each of you knows are appropriate for discourse between airline agents and their clients—a prime example of communicating effectively for the task at hand, which is what good human-machine interfaces should do.

Without a narrow domain, a system like Pegasus would get about 95 percent of each spoken word correctly. Sounds good. However, compounding these errors over five-word sentences would cause Pegasus to correctly understand only about two-thirds of the sentences. Not so good. With the corrections supplied by a narrow domain and language constraints, the system correctly recognizes almost 90 percent of the spoken sentences. This is substantially better. Though it is not the 99.5 percent we humans achieve, it is good enough to be useful.

The presence of such errors in speech-understanding systems is also a good reason to stick with interactive dialog rather than one-way transcription. If you were to dictate a long message detailing where you wanted to fly, what days, what airports you wanted to use, how much baggage you had, the kinds of seats you wanted, who you were traveling with, the times of day you wanted to come and go, the class of tickets you needed, and on and on, and only 90 percent of each sentence matched what you said, it wouldn't be a productive process. You might end up in Des Moines instead of Detroit on Sunday instead of Monday and at 2:00 A.M. instead of 2:00 P.M. But if you stopped after every sentence for a computer nod, you would know when the program misunderstood you, and you could repeat your statement, probably rephrasing it, so the communication would be correct. This is how we communicate with one another, pausing for the slight nod and rephrasing when necessary.

Making good speech-understanding systems requires making systems that have knowledge in a specific domain—a difficult undertaking. This quest has been under way for a long time. In the mid-1970s I tested an ambitious speech recognizer built by pioneer Fred Jelinek when he was at IBM and was trying to construct a memo transcription machine. To define a domain he used thousands of real memos written by IBM executives. After testing the system, I picked up the microphone and let out a long chain of Greek curses, the way we used to launch them at each other in the streets of Athens. The system promptly translated my invectives to "With reference to your memo of . . ." and other such business sentences, because that is all that the system could "understand."

Inspired, I performed a similar experiment on the fellow who sold newspapers to drivers sitting in line at Newton Corner, a Boston suburb. Every Saturday night he would hawk the two Boston Sunday papers, the *Globe* and the *Traveler*. I reasoned that his domain of discourse had been honed to only these two words. I would lower my window and scream, "Obe!" purposefully omitting the first two consonants. He would promptly hand me the *Globe* without question. On another night I would say, "Wawawa." Apparently, the three repetitive syllables were enough to be perfectly understood as *Traveler*. One night, prompted by the demon within me, I shouted, "Obe-wawa," to him. He stared in disbelief and blasted back, "What?!" Computer speech systems are no different. Unless they can find a "slot" of knowledge in which they can fit what they hear, they too will say, "What?"

A speech-understanding system need not be confined forever to a single domain. It can be made nimble enough to change domains through either an explicit command or its own detection that the human is interested in a different topic. This is exactly what happens in Pegasus's successor, Galaxy, which is aimed at helping people visiting a new city. Galaxy combines the airline domain with a weather domain and a city navigation domain, which can look up names and addresses in an electronic phone book and show their locations on a city map. After making your flight reservations, you might ask about the weather in the place you are flying to and perhaps for a satellite image of that area. Without warning you might then pose a question about how to go from Harvard Square to the airport or how many Chinese restaurants there are in Boston (you are shown a map), narrowing it down to the Chinese restaurant closest to MIT. Galaxy was working at this level in 1997 in our laboratory.

Speech-understanding systems with vocabularies of a few thousand words that stick to specific domains, continuous speech, and interactive dialog should be commercially available well within a decade. Galaxy runs on a system that cost $15,000 in 1997. It should be able to run on a PC by 1998 and on a portable navigator device costing a few hundred dollars by the year 2003. Thousands of other commercial applications should materialize by then. In the meantime, we will undoubtedly see systems with more limited vocabularies operating a host of different appliances, like TV sets and washers and dryers, graduating to word processors and programs that will help us speak by phone to computers with our questions on the weather, the news, stock prices, our bank account balances, and eventually many of the activities in the Information Marketplace.

Bodynets and Smart Rooms

The interfaces we've discussed so far have been under development for years. Are these all of the options? Certainly not. Technology is as unlimited as human imagination! Let's visit some strange new breeds of human interfaces that have recently been proposed that integrate different devices.

The first one, called "bodynet," is the brainchild of Olin Shivers at the MIT Laboratory for Computer Science. Before coming to MIT, Olin lived in electronics-happy Hong Kong, where he would constantly see people walking the streets overloaded with gadgets— a laptop computer, a portable radio, a portable TV, a cellular phone, an electronic diary, and a watch. Nearly every one of these gadgets had its own display and keyboard or keypad. The gadgets also had duplicate functions; for example, names and phone numbers were stored in both diary and cellular phone memories. Olin sought a way to integrate all this duplication and confusion.

The Shivers bodynet builds on a pair of "magic glasses" that you wear. They have clear lenses that let you see where you are going but also present miniature inset displays that show color images to each eye. The images are generated by a computer the size of a cigarette pack on your belt or in your purse. The glasses also have photodiode sensors that monitor the whites of your eyes in order to detect where you're looking. Miniature microphones and earphones attached to the glasses let you speak to and hear from your equipment. All the other gadgets that you may want to carry with you share the glasses as their human machine interface and the one computer as their computer, so they can be very small. The cellular phone, the diary, the watch, the TV, and the radio—each smaller than a matchbox—can all easily reside on your belt or on your wrist or in your pocket. You might also wear a ring, preferably on your fourth finger so that you can "click" it with the thumb of the same hand to control these devices. This is no different than the button of a mouse, except that it is where your hand wants to be rather than where the mouse happens to be. The gadgets communicate with one another in a language called "bodytalk," which is transmitted via low-power radio waves that are confined to an invisible envelope around your body—the body network, or bodynet.

You are walking down the street refreshed from your Saturday afternoon basketball game. That hot shower felt good, you are thinking, when someone calls you on the phone. You know this because you see a small, dim green light flashing in the upper righthand corner of your magic glasses. Your "cell phone" is really a software application that utilizes many gadgets on your bodynet. You glance briefly to the right, and the computer, sensing your eye motion, is commanded to answer the phone. It's your mother, chitchatting as always about your daily plans. You hear and talk with her thanks to the small earphones and microphone on the glasses.

Never mind that you appear to be walking down the street talking to yourself. That phenomenon predates the Information Age and is already tolerated by society, especially in large cities. By the way, if you had not wanted to answer the call, you simply would have glanced to the left and your computer would have told your mother that you were not reachable, or as they used to say in Victorian England, "The master is at home, but he is not receiving."

You finish with Mom and continue walking along the avenue, noticing other people talking as you just were. To call someone, you would whisper the name and the call would be initiated. To see the latest television news you would glance up momentarily to turn the TV on and click your ring to change channels. As you walk by one of the high-speed network nodes, housed in what used to be one of those old-fashioned telephone booths, you voice the command for a data refill, and a burst of information is transmitted from the phone booth transponder to your bodynet and computer. The bodynet is equipped to do these transfers automatically whenever you pass by such stations, but you have overridden that capability because you like to control the transfers yourself. You review your personal video messages as you stroll. Suddenly your attention shifts to a well-groomed Airedale that seems to be wandering the street lost. You take out your digital pocket camera and turn it on. It handshakes electronically with the bodynet and becomes part of it. You speak into your glasses, commanding your camera to set itself for the clearest rather than the fastest picture, and then you point it at the dog. The image is immediately stored in your computer.

Encouraging the dog to follow you, you walk toward the phone booth. Your bodynet streams the image data to the booth, which forwards it to the dog registry. There the pictures are automatically identified and routed to the dog's owner, who, alarmed by the news, rushes to the scene a full eleven minutes later. She hugs the Airedale and approaches to thank you profusely for your altruism. As she comes close, her bodynet handshakes with yours, and you are embarrassed to see on your glasses that you are the one with the inferior model. Her system will have to stoop down to your slower speed and coarser video resolution. As you ponder the need to upgrade your system, the two bodynets exchange your personal auto-profiles—part of the routine mutual identification process. You now see in your glasses that though you have no friends in common you share an interest in early music. Seeing a similar message, the woman smiles appreciatively and lets her bodynet burst to yours her favorite Italian madrigal, Orlando di Lasso's "Matona Mia Cara." You are surprised that you do not know this captivating melody. You ask her if she knows what the words mean. She extends her hand for a real handshake and explains with her own sweet human voice that she'd rather wait until she knows you better; they are just too naughty. Your adrenaline surges. You ask her to take a walk in the park. She gives you one of those glances that say more than all the bodynets of the world and accepts. The dog tags along.

Personal networks like the bodynet are also being pursued by Phil Carvey in his Body Lan project at Bolt Beranek & Newman (BBN). The gadgets like the microphones, speakers, video chips, and so on that could be worn on such networks are tackled in many places because they are useful as peripherals for all sorts of applications. Though the bodynet and the streaming of data to phone booths are technically within reach today, the magic glasses will take another five to seven years to become commercially available at acceptable quality. By far, the tougher problem will be to get people to agree to manufacture their various wares for the bodynet.

So far we have talked about an interface that is body-centric: it follows you wherever you go. Now let's escape this confine and consider an interface that would reside all around you. The late Professor Alan Newell of Carnegie Mellon University envisioned this possibility when he wrote in 1976 about an enchanted world in which bridges would watch for the safety of those crossing them and street lights would care for those standing under them. Today, this kind of integrated interface stems from research started by Mark Weiser in 1988 at the Xerox Palo Alto Research Center. Originally called ubiquitous computing, the concept grew to include the "living room of the future" and the "smart room" and was expanded into "things that think" when the work became a focus of the MIT Media Lab, run by my colleague Nicholas Negroponte.

In plain English, ubiquitous computing is an approach that embeds computers in the world that surrounds us. The computers are built into walls, floors, seats, desks, beds, ceilings, kitchen appliances, and lawn mowers, so integrated with their surroundings that you no longer perceive them as computers. Michael Hawley, principal investigator of the project Things That Think, argues that because chips and bits can be easily embedded into nearly everything, we should look at everything that surrounds people as candidates for such interfaces. Bodynets and the ubiquitous computer are complementary: the bodynet moves around with you while the ubiquitous computer resides within the things that are around you.

Having said farewell to your new friend, you walk through your front door and shed your glasses as the bodynet synchronizes itself with your living room computer, transferring data from the info-events of the afternoon. That antenna printed in the fabric of your shirt, despite its one-millimeter thickness, is more annoying than the online clothing catalog in the Cambridge World Shop led you to believe. You change into your well-worn polo shirt. You enter the kitchen and ask out loud for a dinner suggestion. The kitchen computer reviews your intake for the last ten days and the house's food stocks. Taking this into account, it then matches what's left to your likes and dislikes, which it has gleaned from your comments during the last few years. It voices two suggestions: "bruschetta with tomatoes or spaghetti a l'olio et aglio." You issue a verbal order for the former, because the estimated cooking time is reported as only thirteen minutes and you are very hungry. The materials begin to be fetched and prepared by the kitchen unit while you retire to the couch.

More relaxed now, you ask aloud for the less urgent video messages that you elected not to preview while strolling. They appear on a modest-sized video screen that suddenly crystallizes on the far wall. As you electronically page through them, the kitchen beckons you in a gentle voice. The bruschetta is ready. You enter the kitchen to see food being served on *four* plates! You are about to launch into a tirade at the cooking software when the front door calls out that three unannounced guests wish to come in. The video screen that comes to life by the door shows your mother and two sisters waiting on the other side. So that's why she wanted to know earlier if you planned to be home tonight. They apologize falsely but in good humor for not letting you know of their plans to sneak in for dinner. They had called the kitchen and told it to prepare four portions of whatever it was you ordered to cover for them. They wanted to surprise you. You briefly entertain the notion of removing the strong access privileges that you gave your mother to act as your proxy in all matters pertaining to your home machines. But seeing her there with that familiar and loving expression on her face, you push rebellion aside and give her a big hug. Your sisters have already started eating.

Virtual and Augmented Reality

Let's now look at interfaces that surround us and make us feel present in worlds other than the one we are actually in. If a computer feeds us all the information our senses need and if it absorbs convincingly all the effector movements we make, how can we ever tell this interaction apart from the real, physical world it presents to us? This question is a variant of a familiar philosophical quandary: How do we know that the world around us is truly there and not in our heads?

Before trying to answer the age-old question in its newborn form, let's take a step beyond the philosophers of yesterday: Why should the computer be restricted to presenting us with a world similar to the one we know, when it is not impeded by natural laws? Why can't it help us fly over all the buildings of Tokyo, skimming along a few feet above the rooftops? Or over an imaginary city inhabited by spherical people who roll down the street and who have three mouths so they can breathe, eat, and talk all at once? It can, in the world of virtual reality.

The main technical idea behind virtual and augmented reality is that the computer, through the head-tracking helmet and goggle gadgets we discussed earlier, "knows" where you are and in what direction you are looking. It projects to your eyes, ears, and other senses what you would see from your vantage point in the virtual world it has been programmed to simulate—another aspect of the real world or an imaginary, concocted world that exists only inside the computer.

If you are not familiar with virtual reality, let me give you an example of a rudimentary early application, a "ride game" that lets the driver—you—compete in various "real-life" events. One game lets you race in the 1996 Daytona 500. You step into a replica of a regulation race car, which is enclosed in a computer-controlled booth. You put on a special helmet and visor and grab the steering wheel. The inside of your helmet darkens, and a three-dimensional view of the actual racetrack appears before and around you. You press the accelerator gently to pull up to the starting pole alongside the favorites. The green flag drops. You're off. You floor the accelerator and are driven back into your seat. The car vibrates heavily. Out your "windshield" and "rearview mirror" you see actual footage segments filmed from a real car during the real race. You see one driver tailing your bumper. Another racer cuts you off. The roar of your engines is deafening, heat pours in through the dashboard. Suddenly another car spins out in front of you. You jerk the steering wheel to avoid it and are thrown into your lefthand door. The tailgater streaks by. All this happens while you're sitting in a Las Vegas hotel lobby in the middle of winter.

Of course, entertainment is only one possible use of virtual reality. In medicine, as we'll discuss later, a surgeon can wear special glasses to see an MRI scan of a patient's brain superimposed and aligned over the patient's actual skull as he operates. (This is working today in experimental hospital trials.) In design, engineers can study a backhoe from the driver's seat to see if the view is clear or from the point of view of the bucket to see the angles at which it attacks the dirt—all before building it—so that changes can be made on the design rather than on the far more expensive prototype and manufacturing molds. (This, too, works today but with highly stylized and limited forms of images.) Other applications might let you fly through history to experience past wars and major events or through our world to see how people live, play, and work.

The new experiences come in different flavors. *Virtual* reality is when you feel that you are immersed in another place. It can bring distant people in contact with one another, too. If you wear goggles for a 3-D view, and a bodysuit that let's you feel objects and be felt, you can visit your friend or lover a thousand miles away. You will each appear before the other, and you will smile, talk, and touch as if you were in the same room . . . almost. The person your partner sees and feels is often called your "avatar," an incarnation of the real you. Lest you get too excited by virtual reality and its uses, we must point out that we are still very far from having such bodysuits and making your avatar look like a believable "you." We'll address these and other such limitations shortly. Meanwhile, your avatar, whether rendered faithfully or crudely, need not resemble you. In fact, as we'll see in the chapter on pleasure, interesting situations can develop through such transformations.

Augmented reality involves superimposing virtual images on real images, as the brain surgeon did. You could wear your goggles and see an overlaid image of the innards of your washing machine. Then, when you look at the inside of your actual machine, the system, which knows where you are looking via head tracking, can line up this image with the real hardware. When you want to make a repair, the video will show you where to place a screwdriver and wrench and how far to turn them. You comply, and magically the problem is solved. Professor Steven Feiner at Columbia University has done precisely such an experiment with augmented virtual reality for the repair of laser printers. Commercial applications should appear within the next decade.

Another important distinction is between *immersion* and *presence*, sharpened by Professor Andy van Dam, a pioneer of graphics and virtual reality at Brown University. Immersion involves goggles or a huge screen that causes the visual experience to surround you. As anyone who has tried it knows, this is a qualitatively different feeling from watching a small computer screen. Presence is the feeling of being "there," with full believability. Immersion and presence can be independent: People playing Doom or another powerful distributed keyboard game get very wrapped up—they experience presence but without immersion. Conversely, people surrounded by panoramic views of still landscapes will feel immersion without presence. To be fully effective, virtual and augmented reality must involve both.

These sophisticated interfaces suggest another possibility—the *supersensory human* being, reminiscent of Superman. Equipped with eyeglass cameras that can see not only visual images but radar, infrared, and ultraviolet images, ears that can hear conversations and higher- and lower-pitched sounds beyond the normal range, an electronic nose (to be discussed) that can detect all kinds of faint smells, and tactile interfaces that translate environmental stimuli like odorless gases or changes in barometric pressure into pressures on our skin, we may be able to sharpen our conventional senses well beyond their usual acuity. Such a supersensory being might be able to detect dangerous situations ahead of regular people and could be a scout or guide into treacherous settings, including the battlefield. Research is now beginning on such interfaces, which means that commercial units may be seven to ten years away. The technical problems seem manageable, but utility is limited to special situations. Hence their impact may not be widely felt.

Another imaginable, if less appealing, "being" that virtual reality systems could create is an automated politician who would appear on TV as if it were the real person. We would put enough information into our computer to simulate the elected official. Then we would fashion a program that causes that person's image to say a sentence we type to it in a convincing way. Today we have trouble creating a demonstration good enough to fool you into believing you are watching the real person rather than an imitation. But the technology will improve. This capability, if it were to come about (in one or two decades), could be used for good or bad. A politician could spread a lot of clones around and let his trusted staffers answer through them questions posed by the people in an electronic conference, for example. Criminals could use the technique to impersonate you. If harm can be done by reproducing someone's signature, imagine what could be done by reproducing someone's face, speech, and gestures. Fortunately, the security technologies we discuss in chapter 4 can be used to discern the real you from the fake you.

Although such a convincing fake is impossible with video today, it is quite possible, even easy, with photography and still images. As we will discuss in the chapter on pleasure, the altered images can become indistinguishable from real photographs and can bend reality to portray almost any desired effect.

One wonders how far into the future these activities might be, given today's virtual and augmented reality experiments with their heavy glasses, stiff gloves, discontinuous video, and overall sluggishness. A big reason for all this awkwardness is that the simulation of reality requires an incredible amount of computation. As you walk through a virtual space, the computer must calculate what you would have seen if you were there. To do so, it must know where you are and where you are looking. It must then calculate how all the stationary and moving objects and creatures that are supposed to be in this space would look to you from your current position and orientation. Even with the fastest computers, this is a very tough task. That's why objects and creatures tend to be crude caricatures—so that the computer can manage to complete its calculations as you move your head. When it cannot do so, the images you see lag behind schedule, and you may suffer from "simulator sickness"—fatigue and nausea caused by the disorientation you feel when the world around you does not conform to what is expected by your senses.

Some people believe it's only a question of time before machines can simulate everything as faithfully as you perceive reality. If they can already feed something believable to your senses, its an easy step toward fooling you completely. Others, like the philosopher Daniel Dennet of Tufts University, argue that the improvements needed to simulate every grain of sand you feel flowing through your hand at the beach are so fiercely difficult, expensive, and prohibitive in computational terms that they will never come about. Dennet may be right: It takes a lot of reality in the form of computers to simulate even a small corner of virtual reality. Trying to simulate a big enough chunk of virtual reality faithfully enough to fool us completely may require more computers than reality can reasonably supply.

Ultimately, no one knows how far these technologies will go, which is a source of either excitement or fear depending on one's beliefs. No doubt, virtual reality will find its niche in recreation, whereas augmented reality will prove useful in medicine, education, and the many other Information Marketplace services that we discuss in the second part of the book. However, before the humies get all bent out of shape about such techno-assaults, we had better revive some old lessons: The ordinary play in the ordinary theater is a simulation of reality with a track record of a few thousand years. It is crude in terms of its mechanical ability to fool our senses, resorting to fake-looking props and our imagination to transport us to virtual times and places. Yet the theater is very effective, especially when the play is good. It can cause people to interact emotionally through laughter, tears, and that exalted feeling of having reached a higher plateau in understanding human nature. Will virtual reality, even if wildly successful, advance our experiences beyond the theater? Will scenes that appear in magic glasses make people laugh or cry more intensely? Will we reach a more exalted state through the newer technologies?

We have good reason to doubt it in the narrow field of theatrical entertainment. Movies and TV, with all their technical sophistication, did not advance comedy or tragedy in any big way beyond what the theater had already achieved. They did bring the ability to experience a virtual world to many more people, but the emotions they evoked were no more intense. Perhaps virtual reality will turn out to be easier to engage in than the silver or video screen. And it will surpass the theater in causing us to further suspend disbelief and even duck reflexively when virtual objects are thrown at us. But it is not likely to affect humans any more or any less profoundly. A color photograph of some devastating event is no more or less moving than a grainy black-and-white newspaper photo or a written account of the same tragedy. Ultimately, it is the budging of human emotions inside us that counts—something that virtual reality can affect in limited ways. We will revisit this important topic toward the end of the book.

Electronic Noses and Haptic Interfaces

We started our review of human-machine interfaces with speech-understanding systems because they are the most natural to us and are nearest at hand. Bodynets, ubiquitous computing, and virtual and augmented reality will also help us, though further in the future. But computers need not talk exclusively to people. They must also be able to communicate with other computers and machines of all kinds.

There are already many simple input and output devices that connect computers to the physical world. Image scanners "read" printed documents and pictures into a machine. Position sensors tell your car's computer how far you've depressed the gas pedal. Filmless cameras transmit a picture directly into a computer's memory. Humidity sensors stuck in the ground report soil moisture for gardening and agriculture. And so on.

How much more could such interfaces do? Could we have an electronic nose that sniffs undesirable substances like drugs and explosives deep in traffickers' suitcases? There are already versions of these devices used at major airports. Might the concept be extended to all smells? We would put the "nose" on a robot and send it into a hazardous environment like a collapsed coal mine shaft. The nose would sniff the air for explosive gases and report its results, even analyzing what chemicals might have contributed the smells. This would work like speech understanding except that the machine would be fitting smells instead of words into expected slots.

We would then don a remote nose interface that repeats the smells the electronic nose detects to our own nose, sitting safely on our face several miles away. In principle, because smelling involves chemical reactions with the surrounding air, such broad-spectrum remote noses are possible. They would be constructed with a set of perhaps a dozen basic chemicals at the remote site that would report to the computer the components of smell—whether each chemical reacted with the air and how extensive that reaction was—much as a digital camera reports the red, green, and blue components of the colors it sees. These numbers would be shipped to a smell generator near us, where they would tell that machine how much to open the caps on little bottles of smell-producing chemicals. Research prototypes of electronic noses are already underway.

Using interfaces to sense real-world conditions and control biochemical processes like smell opens the door to larger possibilities. Much research in the coming century will explore the relationship of information technology to physics, chemistry, and biology. It's even conceivable that the basic laws of physics, the way chemicals react and biological organisms grow, might be expressed in a handful of "programming rules"—in other words, with pure information. With the right kind of "computer" and associated techniques, so the fantasy goes, we may be able to control physical and biological processes at their very roots. Readers familiar with biotechnology and the mapping of the human genome know that the long journey toward these explorations has already begun. The ethical and technical questions surrounding these quests are monumental and a subject for another book. We mention the topic here because it touches on human-machine interfaces and because it opens, however tentatively, the fascinating possibility that information technology may not play its full hand with the Information Marketplace but may prove to be even more basic to our physical world than its nonphysical nature might suggest.

The ultimate grand discovery in this area would be worthy of a dozen Nobel prizes: it would establish that the entire world around us, all the physical processes, including biological and chemical activity, can be explained in terms of a few programming rules. In other words, information would be the underpinning of all creation! That, of course, is speculation.

In more practical terms, interfaces translate physical positions, movements, color, light, sound, temperatures, smells, and volumes into the 1s and 0s that represent them, and vice versa. These interfaces are the eyes, ears, mouths, arms, and legs of the information infrastructures that will surround us. They will be used increasingly to connect computers to us and to the rest of the world. In the 1970s only 10 percent of a computer's commands dealt with input-output devices like displays, printers, and keyboards. Most commands were devoted to computations that transform information inside the computer. In the mid-1990s that fraction was at 85 percent, and it was rising! In a funny way, it seems that our computer systems are imitating us by devoting increasingly more of their power to input-output tasks: fully three quarters of the human cortex is devoted to vision, the principal input interface of human beings.

The success of speech understanding might suggest that a similar success can be had in vision. And to an extent the analogy holds: Commercial systems (for example, Cognex and Orbotech) work well at inspecting circuit boards for errors, because the domain is small and the systems have a good deal of knowledge about what they expect to see. But a computer cannot tell you that it saw a girl holding a white dog on a leash. We humans can do this instantly because of the wondrous properties of our eyes and brains. Because we do not yet understand these properties, we cannot replicate them for the computer, any more than we can make speech systems understand political and philosophical discussions. Image understanding, however, is more complicated than speech, because it involves one more dimension and complex qualities like textures, shades, hues, and reflections, all of which must be assessed for proper image understanding. It will probably take quite a few more years before image-understanding systems become as widely useful to people as speech-understanding systems.

Nevertheless, vision systems are already important in specialized applications in medicine and manufacturing and are potentially very useful to all people. So research and results continue. In one example, the computer is taught to discern roughly similar scenes in which the skies are comparably blue and the grass comparably green so as to help retrieve pictures from a large picture archive. In another system, being built by Professor Seth Teller of our lab, you can roam around Boston pointing your camera here and there. As you do so, the system identifies and discerns different buildings and structures and constructs inside the computer a 3-D model of the city, which is very different from a collection of snapshots. Later you can reach into the accumulated one terabyte (one million megabytes) of data, pluck out a building like the Government Center, and replace it with a stadium for urban planning and visualization purposes.

In the visual output category of interfaces, we can expect flatter, larger color displays in the next five to seven years, including large "whiteboards" on which people at different locations can view, write, and edit the same document, drawing, or photo. The computer keeps a copy for reference. Early versions of such devices are already being sold.

A fascinating class of interfaces called *haptic* combine manipulation with touch sensing. In one particularly exciting research project, called PHANToM Haptic Interface, J. Kenneth Salisbury Jr. of the MIT Artificial Intelligence Lab has constructed a finger "glove." You put your finger in a "thimble" and move it around. On a computer screen you see an image of your finger mirroring your movements in a space of simple virtual objects like balls and cubes. You "hit" one of them, and miraculously you feel the impact on your finger as the object goes bouncing away. You grab a pucklike object and nudge it along a narrow passage; you see the results on the screen and feel the contact forces on your finger.

Passive gloves with names like VPL DataGlove and the Virtex CyberGlove have been used to monitor hand and arm motions without exerting forces on the user. Early experiments with force-exerting displays have been carried out by A. M. Knoll at AT&T, P. J. Kilpatrick and F. P. Brooks Jr. at the University of North Carolina at Chapel Hill, and Kent Wilson at the University of California at San Diego. The research prototypes will evolve into haptic interfaces—haptic gloves useful in extending the reach of human hands across the Information Marketplace. In the longer term, the approach will probably be extended to bodysuits that can detect your body's motions and transmit touch sensations to your entire body.

Many other human-machine interfaces are either here or around the corner for people who are handicapped. Already, people who are paralyzed benefit from interfaces that can detect the movement of their eyes or tongue and use these cues to control everything from the motion of a wheelchair to the typing of a document. Cochlear implants help some deaf people hear. Prosthetic retinas made out of light-sensitive silicon circuits are likely to be experimentally available within a decade for people who have a damaged retina but a healthy optic nerve. The artificial retina is surgically placed over the optic nerve, where it converts light passing through the eye into tiny electrical signals picked up by the optic nerve, which are sent to the brain and, it is hoped, result in vision.

Many other fanciful interfaces are possible. Sand your stomach with sandpaper (lightly!) to sensitize it and place a special pad over it, the size of a magazine, equipped with tiny electrical actuators that tickle the skin with tiny electrical shocks in a few thousand places, depending on what a camera installed in a helmet you are wearing sees. Then teach your stomach to "see" by learning how to understand the tickling sensations. Or imagine a flat tabletop display in which hundreds of thousands of pin-sized pistons, driven by a computer, rise up to form a relief picture that a blind person can feel—a kind of graphical Braille interface. Such interfaces would be truly useful to people with a variety of impaired functions.

The Ultimate Human-Machine Interface

Today we hear quite a bit of hype about wiring our brains to computers, which would eliminate the need for complex speech processors, video screens, indeed, for all computer interfaces. Proponents argue, "Why not mainline the information directly into and out of the brain, eliminating the middlemen?" It makes great press, but it is not the ultimate human machine interface!

Even if it would someday be possible to convey such higher-level information to the brain—and that is a huge technical "if"—we should not do it. Bringing light impulses to the visual cortex of a blind person would justify such an intrusion, but unnecessarily tapping into the brain is a violation of our bodies, of nature, and, for many, of God's design. For those impatient with ethical and religious arguments, let me suggest some practical reasons: A certain amount of isolation among interacting beings is essential for the orderly functioning of the aggregate. If people were connected via their brains rather than by slower, more isolating interfaces, they would be flooding one another with messages as fast as they were thinking them up, would hardly have any time to think for themselves, and would be acting more like a single (confused) organism than a group of independent agents. In fact, the very distinction between a single organism and a society of organisms seems to hinge on a goodly amount of separation. Fear of unknown consequences is another practical and proven reason for avoiding casual implants, as we discuss in chapter 7.

Coming back to today's world, what we hear touted almost constantly is the word *multimedia*, which many suppose to be the ultimate new computer interface, full of dynamism and user interaction. Well, such hoopla might impress somebody's senses, but by itself the notion of multimedia is neither meaningful nor useful.

Strictly speaking, multimedia means that your computer will use text, pictures, sound, and video in the same application. Though it is intriguing to have these different kinds of information, it is not meaningful unless they do indeed work together and enable you to do something better than you could have done it before; otherwise, it's like a person who collects many different tools but cannot use them together. This is largely the state of multimedia today.

The promise of multimedia includes a lot of wishful thinking. Computers cannot process video images the way they can text, so it's difficult to make these two media truly work together. Show me a computer that can look at a video and identify that a little girl with a white dog on a leash is running across a street and is about to get run over by a BMW convertible, and I'll show you a fake! With text, computers can easily identify the component words. One reason is that pictures are represented with pixels, which do not reveal easily the picture's "components." Of course, things get worse if instead of a single picture we are dealing with hundreds of thousands of pictures and the movements they paint in a video.

Why am I throwing this tantrum? Because the mindless use of the term *multimedia* occludes the real problems and the true excitement in human-machine interfaces. For example, it would be useful to have the different communication modes reinforce one another for a single purpose. To edit text, you might simply use a stylus and strike a line through a word on your screen as you simultaneously say *delete*. Here, the use of two media—speech and image—serves the useful purpose of conveying an unambiguous message—like shouting, *"Come over here!"* while jerking your head back and gesturing with your hand and curled index finger. Either method used by itself can convey the instruction, but the three taken together make for a far more powerful message that's harder to misinterpret. Interfaces that gang different sensory and effector modalities together to convey a single message are called *multimodal*. They can make multiple media work in concert for us.

As powerful as the multimodal notion is, it is not always the preferred approach. Signing a contract places a premium on being able to read the written language to be agreed upon. Babbling away and gesturing compromises the process. In certain teleconferencing experiments, shutting the video down and leaving the sound on improves the participants' focus, because they are not distracted by the video. Some people prefer getting certain kinds of information from a book rather than a movie.

To complicate matters, cognitive scientists like John Seely Brown, head of Xerox Palo Alto Research Center, tell us that the important action in the office happens near the water cooler where people chat. Attending a conference, they also say, often has little to do with listening to the speakers; people want to network during the breaks. As a result, Seely Brown claims that we must shift the focus of human computer interaction to designing the social and informational periphery as opposed to just its center. We need to take these important observations into consideration when designing effective human interfaces for work at a distance.

Tomorrow's human-machine interfaces will also improve as we learn how to pull them away from the drudgery of machine details to higher conceptual levels, closer to the way we think. The idea is like teaching a busboy in detail how to clean the tables in a restaurant and then later simply telling him to go clean the tables without restating all the details again. This move up to a higher level of abstraction is a powerful tool for increasing human productivity. The same move will happen in human-machine interfaces; instead of doing everything ourselves by pointing a mouse, clicking, typing, and even issuing detailed verbal commands, we will encapsulate a set of instructions to the machine into a single command—one mouse click or typed word or spoken phrase. The machine will do the rest, helping us greatly.

These observations lead us to conclude that the "ultimate" human-machine interface is the one that brings together the right modes of communication, the right hardware, and the right software—all tailored to the kind of concept that is to be conveyed between human and machine. Ultimately, it does not matter how many displays are embedded in the walls of your house, or how many gadgets will hang on the belt of your bodynet, or whether speech will be used more or less than keyboards and mice, or whether you can fly through your data and see virtual animals floating in space as you execute your commands, or whether you'll use one or six media. What really matters is that you, the human, are trying to send or absorb a concept. All this fancy machinery had better help you do exactly that. The central purpose of the ultimate human-machine interface is to achieve communication on human terms—what we have come to call language in its broadest meaning. When this happens, effectiveness will supersede pizzazz and drive the development and use of the best interfaces for the multitude of tasks that will take place in the Information Marketplace.

Techies of the twenty-first century should focus their attention on understanding and developing human interfaces that can provide the most effective communication of different classes of concepts as executives, surgeons, engineers, artists, teachers, and all sorts of different people try to communicate their indigenous wishes and tasks to their machines—and the other way around. Sometimes a lot of gadgetry will do it best. Sometimes less will be more: Imagine that you had in front of your computer screen no mice, no speech processors, no virtual reality, and no multimedia, but a simple keyboard with the iron-clad guarantee that the computer would understand everything you typed as if it were a human. Would you like that imaginary interface, dull as it may be, or would you opt for the fanciest multimedia interfaces on the most powerful supercomputer that money could buy today but with their current level of understanding?

Tomorrow's human-machine interfaces will vary in form and function, and they will often reinforce one another. They will operate at higher levels, closer to the way people think and interact naturally—and, together with the software that drives them, will be tailored to communicate something specific. What are these "somethings" flowing between us humans and the machines around us? And what is it that we expect our machines to do in concert with all the other interconnected machines after we communicate with them? On to the heart of the Information Marketplace—the tools we will use to buy, sell, and freely exchange all that information and those information services.

4
New Tools

The New Software

Human-machine interfaces will give you access to the Information Marketplace. But once you are in, you'll need tools to help you manage your buying, selling, and free exchange of information and information services. There are some big promises in this book about activities you will pursue on the Information Marketplace: shop, play, work, improve your health, conduct business, teach your children, preserve your heritage, control access to your personal records. But how will you actually *do* these things? With new tools. We've looked at the interfaces of the future; it's now time to look at the tools of the future, especially those that will be widely shared.

Some people claim the tools we need are largely here. Not so! With today's computer networks, you and your computer applications must labor hard to accomplish something useful among interconnected machines. With tomorrow's information infrastructure tools, the computers will do more of this work for you. The difference is huge. For example, with current network tools you can't easily bill for services rendered or pay for services received. You can't work together effectively with other people. You can't automate routine transactions among machines. And you can't organize information, even though there is a great deal of it out there. The things you can do, such as edit documents or coordinate calendars, often can't be shared with others because different systems and service providers all have their own internal mechanisms for doing these things—which they don't share. As a result, on today's networks it's rare to be able to cross from one application to another, or from one service to another, except for sending e-mail or clicking on a Web page—and you are doing all the work because you must either read or write and interpret what you see to determine what you should do with it.

This situation is not increasing our productivity. On the contrary, we must stand on our heads to develop piecemeal solutions that substitute for all the tools we do not have. Though tomorrow's infrastructures may evolve along different paths, or "architectures," they will nevertheless end up having certain basic capabilities that will be shared by all applications. These are the useful tools we'll examine here.

The widely shared software tools, sometimes called "middleware modules," will be used like building blocks by people and computer applications to construct desired functions and services. The script specified by the Philadelphia doctor to handle X rays would be carried out by a pipe manager—a middleware module that helps transport information. Another tool, the hospital's billing middleware module, would handle charges for Dr. Kane's services to Ruby Creek, along with his billing for internal patients. These modules might be stacked together automatically by yet another middleware tool, so all the computerized activities would be coordinated and take place automatically. Or they might be specified directly by the people using them through simple scripts. For more complex tasks, these shared middleware modules would be artfully interconnected by specialists—the information infrastructure programmers.

Today's "shared" tools, offered by the network information services and the World Wide Web browsers, are front-ends that allow you to perform certain fixed functions such as accessing stock prices, getting the news, or placing an order. They aren't evaluating the price changes or filling out the order forms for you, or doing any of a thousand other chores you would like done.

Like today's computer programs, the new tools will be numerous, they will be offered by many suppliers, and they will vary dramatically in their quality and cost. The twenty-first century will see entirely new software industries created to cater to this new genre of programming. Initially, widely used middleware modules like stock analyzers and automatic order-form completers will be offered. So will specialized products: doctors will have their own groupwork modules catering to their needs, salespeople will have theirs, and so on across the entire economy. This software will be developed and sold by the independent software vendors of the world, who today produce some 15,000 different shrink-wrapped software products for standalone computers. The new tools of the Information Marketplace will be their new challenge—and their grand opportunity to expand the value they offer and the sales they make.

Unique in-house software tools will also be developed by the world's organizations. In commerce they will become protected strategic assets, enabling companies to exploit their special advantages ahead of their competitors. Individuals will program middleware too, just as some people program their standalone machines today. Though personal software development is waning today because so many programs are commercially available, this trend too will change, as we discuss in chapter 12, surprising us with its vigor and consequences.

The distinction between the dedicated applications that you and I will use and middleware software is fuzzy. In fact, as we'll soon see, an important aspect of an information infrastructure is that today's fledgling applications may become tomorrow's shared tools.

No one can know what useful software might emerge in coming decades. But today the first true middleware modules are in the offing. There are half a dozen categories of modules poised to become major, shared tools: automatization, e-mail, groupwork and telework, pipe managers, hyper-organizers and finders, and computer security and payment schemes. By looking at them now, we can learn a good deal about how we and our machines will do what we want to get done on the Information Marketplace every time we use it.

Automatization Tools

One of the biggest roadblocks to building an Information Marketplace is the inability of interconnected computer systems to easily relieve us of human work. This roadblock exists because today's networked computer systems have no way of understanding one another, even at a rudimentary level, in order to carry out routine transactions among themselves. It isn't simply that IBM machines are different from Apple machines, or that the Windows operating system differs from Unix. The problems run much deeper: to the absence of shared concepts and capabilities even among an army of identical IBM PCs all running the same version of Windows and all having exactly the same applications. The problem is that we know well only how to harness a single computer rather than a whole bunch of cooperating machines. Add the differences among the hardware and software products of different vendors, and the problem becomes even harder to solve.

To offload human brain work, an information infrastructure must ensure that its interconnected computers "understand" enough about one another to work together. We call the new tools that will make this possible "automatization" tools to distinguish them from the automation tools of the Industrial Revolution that offloaded human muscle work.

Today, we are so excited by e-mail and the Web that we plunge in with all our energy to explore the new frontier. If we stop and reflect for a moment, however, we will realize that human productivity will not be enhanced if we continue to use our eyes and brains to navigate through this maze and understand the messages sent from one computer to another. Surely we don't need computers forcing our brains to read textual communications. Telegraphy accomplished the same objective a century ago! Yet this is what goes on today with 95 percent of the world's computer network activities, under modern names like e-mail, newsgroups, bulletin boards, and browsers!

Imagine what would have happened if the companies making the first steam and internal combustion engines of the Industrial Revolution had made them so that they could work together only if people stood between them and continued to labor with their shovels and horse-drawn plows. What an absurd constraint! Yet that is what we do today—expend a huge amount of human brainwork to make our computers work together. It's time to shed the high-tech shovels with the fancy names and build the bulldozers of the Information Age—with or without fancy names! That's what the automatization tools are all about.

Someone familiar with the cutting edge of Internet work may now jump up and say, "Aha, that's what Java and languages like it are all about." Not quite! Java and the others can be used to program useful automatization tools, but the languages themselves are not enough to achieve automatization. When the landmark programming language Fortran was invented nearly four decades ago, spreadsheets didn't automatically follow. They had to be created from whole cloth, many years later, and then programmed using languages like Fortran. Moreover, practical programs like Excel and Lotus 1-2-3 had to be developed beyond that point and at great cost before spreadsheets actually performed useful work for us. Similarly, new tools will have to be invented, programmed in Java or some other language, and packaged into robust products before practical automatization among machines becomes truly possible and widespread. That is where the hard problems lie, and that's where techies should focus their creative thrusts.

Achieving some basic degree of understanding among different computers to make automatization possible is not as technically difficult as it sounds. However, it does require one very difficult commodity: human consensus.

Let's look at a brief example to show what it takes for machines to reach agreement on a simple task. One easy way is for us to agree on the meaning of certain nouns and prepositions that computers exchange with one another. Take the words *number, date, from, to,* and *class* as they are used in airline reservations. Suppose also that software makers agree on how to cluster these words together in a simple tool we'll call an "electronic form," or "e-form."

The e-form

number	2
date	25.08.96
from	Boston
to	Athens

would be understood to mean a flight reservation fragment.

Let's now add to the above words a few commonly agreed-upon actions or verbs, like *available, understood, book,* and *confirmed.* And let's agree that these verbs will be normally used as affirmative statements, or as queries when followed by a question mark, or as negative statements when preceded by the word *not.*

We can now imagine the following dialogue between my computer, on the left side of the page, and the airline reservation computer on the right side:

Available?

number	2
date	25.08.96
from	Boston
to	Athens

NOT AVAILABLE

Available?

date	26.08.96

NOT AVAILABLE

Available?

date	27.08.96

CLASS?

Business

AVAILABLE

Book

UNDERSTOOD AND CONFIRMED
CERTIFIED PAYMENT DUE ONLINE BY 20/08/96

My machine, hearing me say, "Take us to Athens next Sunday, Monday, or Tuesday," would carry out this entire dialog on its own while I was checking the calendar of events in Athens during late August using Galaxy.

My personal flight reservation program already knows that unless I specify otherwise, the journey will originate in Boston, it will involve business class, and that "us" is two people. It also knows how to connect itself to the airline computer, hold the above negotiation, disconnect, and present me with the results.

With this automatization tool, it would have taken me a mere 3 seconds to orally convey this high-level order. Otherwise, I would have spent at least 10 minutes connecting manually to the airline reservation computer and waiting between my keystrokes for the inevitable round-trip delays typical of these systems. I was able to do the job in a tiny fraction of the time. I could rightfully brag that my productivity gain was 200 to 1 (600 seconds down to 3 seconds), or 20,000 percent!

The details of this example are not important. The point is that agreement is needed on the shared meaning to be attached to words that machines will use before useful work can be done. E-forms are not the only way to achieve automatization. There are many ways to construct computer-to-computer languages, with their associated meaning regimes, even ones that look almost like English. Furthermore, it is likely that on any information infrastructure there will be several such languages catering to different specialties, just as we have different sets of words for accounting and gardening. In fact, the automatization tools for such dialogues will have to cope with a familiar problem—words that have different meanings in different contexts, and among different computer communities, such as the word *post* in accounting and gardening. Some of these infrastructure languages and commands will be very narrow and confined to their specialties, while others like *yes, no, do you understand?* or *do you have xxx?* will be so widely shared or mutually inter-translated that they will gradually form the infrastructure's shared-meaning base—call it "computer English"—which along with its various dialects may someday evolve and be widely used in the Information Marketplace.

Thus we can imagine that in the Information Marketplace, common interest groups will establish e-forms to specify routine and frequently recurring transactions in their specialty. If the members of such a group can agree on an e-form, especially one that represents a laborious transaction like spelling out the specifications needed to control and route an X ray, then they will have achieved substantial automatization gains. Computer programs or people interested in doing that kind of business would be able to look up the agreed-upon e-form and use it in their computer, toward the same gains but with much less effort.

Techie readers familiar with artificial intelligence should not overestimate what we mean in the above discussion when we say that machines must understand shared concepts. A long path seems to lie from the shared concept of an airline reservation to the shared concept of compassion. The concepts that computers will handle in the foreseeable future will be closer to airline reservations!

Quite a few computer wizards, and people who are averse to standards, believe that common conventions like shared languages and e-forms resemble Esperanto, the ill-fated attempt to create a universal spoken language among all people. They argue that attempts at shared computer languages will suffer the same ills. Instead, they say that the only way our computers will get to understand one another will be by translating locally understandable commands and questions among their different worlds, just as people translate between English and French.

This argument is faulty, because shared concepts are required even in translating among different languages. Whether you call an object *chair* or *chaise*, it is still the thing with four legs on which people sit. It is that shared concept base, etched somehow in your brain, that makes possible a common understanding of the two different words in English and French. Without it, no amount of interconversion can lead to comprehension, simply because there is nothing in common on either side to be comprehended!

As we can form a consensus within and across specialties concerning the basic meanings that computers share, then even if we end up with different languages and dialects, software developers will be able to write programs and ordinary users will be able to write scripts that install useful computer-to-computer automatization activities—searching for information on our behalf, watching out for events of interest to us, carrying out transactions for us, and much more.

As the Nobel laureate Herb Simon has noted in his book *Science of the Artificial*, disciplines like mathematics and computer science do not have natural laws as do physics and biology. The burden is upon us, the makers of these artificial systems, to define the laws—

in this case, the concepts that should be shared—so as to achieve utility and orderly growth. This is an area where computer technologists should focus their research if we are to make headway in this important business of offloading human brainwork on machines.

Programs that carry out actions on our behalf are sometimes called *knowbots,* a word coined by Robert Kahn of TCP-IP fame to parallel the notion of robots. They are also called *agents,* or *intelligent agents,* evoking the image of able representatives that purposefully carry out our work. Good progress is under way. For example, Patti Maes of the Media Lab at MIT has created agents that indicate their progress with icons on your screen that show different facial expressions. They smile if they are meeting their goals, look sour if they are failing, and look confused if they don't know what's happening.

The prospect of an electronic agent that roams around the networks and does all sorts of impressive things on our behalf has been excessively hyped, pumping up our expectation that these programs will be able to do things we don't know how to do, in some mystical and magical way. If a purported agent can share certain meanings with the sites and agents it visits and if you can understand, as you do with other conventional programs, how and what the agent does with these shared concepts, then you are dealing with a genuine automatization tool. Otherwise it's most probably wishful thinking or worse!

Enhanced human productivity is not the only benefit of the new world of information. Other notions we value—quality of life, convenience, access to knowledge, peace of mind, and better human relationships—will also be affected profoundly by the Information Marketplace. But productivity enhancement is so central to freeing people from work, and it has been so crucial to the success of the two earlier revolutions, that it stands our as a key factor to be exploited, perhaps the key benefit ahead. Yet, it seems to be either disbelieved or aggressively ignored by the media, pipe, hardware, and software companies swarming around the Information Marketplace. I know of no such company developing the mechanisms or the substance of shared concepts toward automatization. Maybe they consider the effort too big, outside their individual specialty, or futile. They should get together with one another and with user groups, as we discuss in chapter 9, to develop conventions for e-forms, for posing and answering queries, for shared nouns and shared actions on these nouns, much as we did in our airline reservation example. The sooner these companies and user groups tackle this need, the sooner they and we will be able to reap the benefits of automatization and increased productivity.

Good Ole (and New) E-Mail

The second major shared tool of the Information Marketplace is rooted thirty years in the past.

The first sign that someone has entered the Internet and Web world is the proud and joyful act of sending electronic mail to friends and getting back answers. E-mail is a basic function on all information infrastructures. It allows us to readily send text, images, and sounds to one another. And it will continue to be a major tool of the Information Marketplace because it is an indispensable mechanism for moving information around. However, its advantages and disadvantages will become much more pronounced as the Information Marketplace becomes more populated. Future e-mail developers and users must take these complications into account as they build the next generations of this useful shared tool.

To better understand what future e-mail tools will have to contend with, and what they might provide, let's visit some extremes.

I have always had an aversion to e-mail. It may well have started in 1974 when we gave e-mail to all three hundred students taking MIT's introductory computer science course, which I taught. The new medium was exciting and even worked smoothly. But it was sheer terror if the lecturer (me) made an error. A few hundred e-mail messages would land in my computer mailbox within an hour. Each student, besides offering a creative and juicy tidbit of criticism, had a special cure to suggest, and each student expected a prompt, personalized response.

A few years later, when the academic community was networked together via Arpanet, the pattern reappeared on a larger scale. My colleagues and I would get a ream of e-mail messages whenever a major blooper happened, or was even *perceived* to have happened. E-mailers had an uncontrollable need to flame (send typed tantrums) in our direction, to vent their frustration and seek corrective action just as our students had. These patterns continue today, when hundreds and sometimes thousands of Internet users go up in electronic arms before a common "enemy."

After a few too many of these e-mail avalanches, I became exasperated and let it be known that, because of its lousy format, I would no longer use e-mail; you had to spend a good deal of time to find out who wrote a given message, to whom the message was going besides yourself, and whether you were the recipient of an original or a lowly 1-in-10,000 copy. This good-natured disinformation worked for a while, until people caught on that I could not ignore e-mail and did in fact read it.

Steve Ward, who would later pioneer the workstation concept at MIT, started using e-mail a lot more enthusiastically. He got equally exasperated when his own barrage of messages started exceeding a hundred per day. But he went a step further. In the early 1980s he designed an automatic filter program that examined a number of criteria like sender's name, number of people being copied, and presence of important keywords about funding or common acquaintances. The filter used these criteria to calculate a number that reflected the "importance" of the message. If the number was above a certain level, it turned the mail over to Steve. The other unlucky souls would get this message:

> Professor Ward regrets that he will be unable to see your e-mail message, which scored 37 out of a possible 100 on a heuristic scale of importance applied by his mail filtering program. If your message is important, please call him at 617-253-xxxx.

When Steve conceived of this hack, neither he nor the rest of us had any idea of the apologizing and backpedaling he would have to do. People were outraged that a mere machine was rejecting them when they had expended a good amount of their own human effort to compose their messages. When this software reappears as an e-mail agent, as it surely will, you can't say that you haven't been warned!

You may think that these stories are snobbish and apply only to people with some public visibility. Not so: As millions of people become interconnected in the Information Marketplace, each of them—each of you—will become much more visible too. Your situation will be no different than mine was during these older experiments. How would you like to receive 1,200 telemarketing e-mail messages a day, all tailored to your interests? If you are using the World Wide Web, you know what I am driving at, because you are undoubtedly already getting unsolicited messages. And please do not look for an "intelligent" agent to fend off these nuisance invasions, because unfortunately there is no program today or in the foreseeable future that can distinguish between a clever telemarketer and someone you might want to hear from.

We'll address the world of info-junk and offensive mail later in the book. For now, let us simply observe that opening your door to e-mail is not far from opening the front door of your home and shouting to everyone, "Come right in, when you wish and as you wish! I will see you and hear you out!" Info-junk is not like physical junk mail. If the mailman leaves an undesirable catalog or envelope full of ads, a simple glance tells you that you can toss it. But you have to read enough of an e-mail message to understand what it is about before you can decide to toss it.

At the other extreme, e-mail can be a godsend. The person who is lonely now has a magical door that opens to thousands of relationships with people who share the same interests or have the same need for companionship. In the mid-1990s tens of thousands of seniors were using the Internet to communicate with distant friends, relatives, and other seniors. Unknown thousands of rural, handicapped, infirm, and homebound individuals were following the same course, as were numerous romance seekers.

We can conclude that for most people who are somewhere between the poles of popularity and loneliness, e-mail pits the power to augment human relationships against the costs of invaded privacy. Future e-mail tools will provide controls that let us receive mail from friends while refusing to receive messages from designated people, anonymous people, or anyone else, for that matter.

E-mail already augments relationships by supporting instant exchanges of sound messages, photographs, and video segments among Internet users. For many, these exchanges are a vivid improvement over the typed letter and its built-in transportation delays. Because e-mail transports numbers, it can carry all kinds of information—text, sound, images, and video but also programs and all the input and output signals of fancy interfaces. So if I had a particularly exciting virtual reality experience with tactile, visual, audio, and olfactory sensations, I could just "e-mail the experience" to you. On your end, you would don your virtual reality gear, feed it my information, and partake in exactly the same sensations I had—for example, the Olympics. What's more, I could replicate and send this experience to many people. It may be ridiculous to wear such a bodysuit while proposing to your sweetheart. But your children and grandchildren would then be able to experience the romantic event responsible for their existence. Rudimentary exchanges like these involving goggles and sound should be available within a decade. The rest could take a couple of decades to come about and would still have limitations depending on developments in haptic systems.

Future e-mail software tools will come in many different styles from many vendors. And like physical mail, they will bring along some unwanted goods: the unsigned letter, the obscene message, the threatening note, the forged dictum. We'll examine some of these issues toward the end of the book when we ask which human activities fare well and which ones do not as they pass through the Information Marketplace.

E-mail may be thirty years old, but it is a basic shared tool that will continue to survive and improve in tomorrow's Information Marketplace for the simple but powerful reason that it can transport every kind of information from anywhere to everywhere. Let's now extend this powerful capability to the information generated and consumed by information workers. The resultant coupling of human information work across space and time is a new, exciting, and promising capability with its own brand of middleware tools—the topic of the next section.

Groupwork and Telework

You are a surgeon at Massachusetts General Hospital who wants to have a couple of experienced associates monitor a critical heart surgery you are about to perform. You step into the operating room and turn on the groupwork software module called ORC (for Operating Room Consultation). Your associates, one at Texas Medical Center in Houston and the other at Mt. Sinai Hospital in New York City, each sit in front of a video screen at their offices, equipped with a conference-type audio link and controls for their own high-resolution video camera suspended from the ceiling in your operating room. They can steer their camera to observe and zoom in on any part of the procedure that you are performing. They can also manipulate laser pointers to point at a precise location when they say "put a suture here." The ORC middleware ensures that all these capabilities mesh smoothly. It has many features, some apparent to you and others not, like the audio silencing, camera retract, and laser shutdown that it performs automatically on loss of communication so that you are not distracted from your delicate procedures.

Maybe you are a New York City detective chasing a gang of criminals shortly after a great robbery at the Metropolitan Museum of Art. You need the help of detectives in Mexico and Canada because the thieves are likely to try to cross the border. You use a CM (collaborative map) groupwork module that displays a common map. The module lets each of you and others involved in the hunt enter information on the map, point to border crossings, mark likely ones, and forcibly move the markers of the other detectives in order to explain a desired course of action.

Or you might be an industrial designer, working from your studio in Milan with a few engineers in a Japanese factory to create the illustrated manual for a newly manufactured house-cleaning robot. The STC (space-time collaboration) module that serves you knows how to handle delayed messages, because the time-zone differences make it very difficult for you all to work at the same time. The module shows and time-stamps each person's editing of the manual's text and redrawing of a diagram and delivers audio sound-bites of you explaining your rationale for your changes. It also keeps orderly records of intermediate and discarded text, pictures, verbal instructions, and "exploded" diagrams. Work advances nearly around the clock without a live conversation.

Routine use of *groupwork* and *telework* modules like these should be possible very soon. Except for the widespread availability of high-speed communications, all aspects of these scenarios are feasible today.

Groupwork and telework modules, which allow several people in different locations to work on a task simultaneously or to build upon one another's work by giving input at different times, could become the most useful tools of the Information Marketplace. They will provide a new and valuable function—linking people together effectively regardless of where they happen to be and when they can be reached. Such bridging of space and time will become more important as people increasingly work with distant partners on an ever-shrinking planet.

Groupwork modules will be used heavily in business for putting together distributed task forces, for carrying out negotiations between people for holding meetings, and for much more. But modules will also be developed for recreation, virtual travel, interest-group encounters, computer romance, and games that will be most exciting when played by large numbers of people.

It will be fun to see where the early modules lead. Remember the group game Maze Wars that I and my colleagues played in the 1970s on our time-shared computer? Well, in the early 1990s, the U.S. Department of Defense built an enviable upscale version of this game called Simnet for tank battle training. The players, who are tank commanders, operate inside a mockup of the interior of a tank. Their view of the battlefield is a computer-generated display. But in a devilish mix of reality with virtuality, some commanders on Simnet are actually riding in real tanks out on the desert, miles away. The commanders on the tank simulators do not know if the tanks they see on their monitors are real or virtual. This war game has been a great success. Like its grandfather, Maze Wars, Simnet is addictive. And even though it is in every way a serious military tool, it is, thankfully, still subject to some playfulness: "Gotcha, Smitty," roars one of the commanders while firing a salvo of shells at his old schoolmate.

Groupwork and telework software will come in many different flavors. To get the full picture, imagine a table with several rows and columns. The rows are the different sectors of the economy, like agriculture, manufacturing, retail trade, health-related services, real estate, financial services, and government. The columns are the various types of groupwork and telework that can be conducted, like conferences, workshops, small meetings, review sessions, and collaborative map sessions. In each square on the table we can imagine several new software tools that would enable that kind of groupwork or telework for that sector of the economy. Doctors' meetings will be supported by different groupwork software than engineers' meetings, which in turn will need different software than engineers' conferences. These thousands of groupwork and telework tools will keep the programmers of the independent software vendor companies very busy indeed.

It's too soon to tell whether and to what extent we will accept all these modes of work, so we can't be sure of their success. However, we can contemplate our familiar mind-boggling statistic: some 50 percent of the workforce in industrial nations works in offices. Just think about how much of that work could be done remotely. We'll come back to the implications of this incredible vision in the third part of this book.

Pipe Managers

In our discussion of the pipe wars, we saw how the telephone, cable, and satellite carriers, as well as some unexpected entrants, might compete for the lucrative business of transporting information. Regardless of how these wars are resolved, it is certain that we consumers will be pursued by competing carriers, each offering a multitude of transport capabilities. How will we ever decide which road to take and which toll is worth paying?

The "pipe manager" middleware modules will help us. This type of shared tool can be thought of as a box with a few levers that control the speed, security, and reliability of information transport that you want. It will have the equivalent of a taxi meter on the side showing how much various services are bidding to do what the levers call for. You can set the levers, or a computer program can do it for you, or you can work together to get the best deal.

The pipe managers will be an important new tool for every information infrastructure because they deal with the essential task of transporting information, providing for an effective match between the growing supply and demand for transport services. To see how one pipe manager might work, let's recall the Ruby Creek X-ray procedure initiated by Dr. Kane. His orders were:

> Send chest X ray to A. Smith at Medlab1.
> Max transport time 2 minutes.
> Min security level is telephonic.
> Min overall reliability 99.98 percent.
> Return reading to M. Kane at Philadelphia General.

The pipe manager would break this command into more elementary parts, for example, carrying the X ray from Ruby Creek to the radiologist A. Smith, and then from Smith's computer to Kane's. The pipe manager would then take the requirements for speed, security, and reliability and negotiate automatically with competing transport services. One cable service responding automatically to this query might report that it can indeed handle all three requirements and can even ship the X ray five times faster than requested, for a flat fee of $1.20 per X ray. A telephone carrier might report that it can handle all requirements except speed, which will have to be half as fast, but at a price of $0.85 per X ray.

For standing procedures like the transmission of X rays, Dr. Kane or his technician or a hospital administrator would have examined the basic options and decided on a long-term trade-off between speed and cost, perhaps buying a wholesale transport service for all the hospital's needs. However, if the transport companies were to change their rates monthly or even hourly, the hospital administration may prefer to program the pipe manager to automatically decide on the best combination for each case. They might specify, for example, that a service be chosen that minimizes the dollar cost per X ray by trading off each second of transport delay against a $0.10 cost savings.

Information carriers will develop many transport capabilities and pricing strategies. Some pipe companies may offer flat rates. Others may price their service at wholesale rates for bulk moves. Yet others may try to capture business more permanently by offering attractive annual rates while their competitors go after spot prices, negotiated by the second. Whatever happens, and there is no telling how such pricing will evolve, there will have to be tools in people's computers where these differences are settled and transport services are chosen and told what to do. These tools will be the pipe managers.

Work is in progress at many places to inject this kind of capability into the Internet. From standards committees to telephone company strategists, everyone is busy trying to divine approaches that offer flexible transport services and pricing mechanisms.

Hyper-Organizers and Finders

An important meeting on the future of global oil supplies is under way at multinational Olie Energetics BV in the port city of Rotterdam, the Netherlands. You are a twenty-first-century secretary charged with the task of summarizing what the five executives and two consultants are saying and displaying around you so that they can later review key portions of the meeting and so that CFO Wilhelmina Maas, who had to be away at one of the firm's drilling rigs in the North Sea, can later find out what was discussed. Any employee cleared to do so should be able to query your notes without having to read through or view the entire two-hour proceedings. You build your summary in a structured hyper-outline form, made possible by the ATM (Authoring Tool for Meetings) middleware module.

As people speak, you hit different keys on your computer keyboard to record pivotal spoken statements or to index something that was said under one of several categories of discussion that you have already set up. You also direct some of the spoken fragments to a speech-understanding program, where they will be transcribed and indexed automatically. You do all this rather well. After all, you are a specialist in hyper-organizing live-meeting notes, and you have spent two years learning and refining the techniques that have landed you this job.

Upon her return late that night, Maas calls up your hyper-summary on her computer and asks, "What did Jan say about recent developments in France's nuclear power supplies?" Maas is rewarded with a couple of sentences answering her question and two pointers—one to an audio fragment of Jan's key statements on the topic, the other to an online text version of the same. If Maas gets her answer in two minutes, and the meeting lasted two hours, then you could rightfully brag that your hyper-summary gave her a 60-to-1 leverage, or 6,000 percent efficiency. Bravo!

Such a tool operated by a skilled secretary is attractive because it helps us zero right in on the information we want and because it does not disrupt the information generation process. If the energy meeting participants had to constantly earmark their own comments or be asked to stop by the secretary, it would destroy the fabric of the meeting.

"Hyper-organizing" and related tools will play increasingly important roles in the Information Marketplace for all kinds of different meetings, for talks and presentations, and generally for key events likely to be revisited after they have happened. Businesses will not be the only beneficiaries. If in January 2001 you are halfway around the globe and can't attend your son's three short appearances in his sixth-grade class pageant, you will still be able to see his grand entrances when you get home by querying the hyper-summary.

We'll also want to use hyper-organizers to manage the text, graphics, video, and audio we get from the Information Marketplace. But before we can organize it, we have to locate it all. For that we will use shared tools dedicated to finding information.

In chapter 2 we saw how search engines accumulate information from the Web by storing all the words they find as they crawl over every Web site they can reach. The resulting index of key words to sites is not very useful, because the search engine will give you a huge rock pile of stuff that you have to manually sift through to find the few nuggets you may want. As time passes, however, this process will improve. Human editors and software tools will help organize these gigantic indexes according to the meaning of their contents, leading to much more useful search results.

After they have been exposed to the Web, some organization freaks invariably suggest that users of the Information Marketplace should organize information as they generate it, according to certain preset rules. This is fruitless because people won't bother to comply. To prove this, just check with people who use the popular Microsoft Word word processing program. Under certain conditions, when they file their document away, a benign form appears on the screen asking them to provide a couple of words of summary information. I have asked hundreds of Word users, and I have yet to find one person who fills out this form. They just tap the Delete key to get rid of it.

This resistance to front-end taxonomies is a consequence of human nature. It is also a valued characteristic of the Internet and the Web's unordered egalitarianism. Any attempt to enforce organizational rules will surely cause users to flee unless the front-end organization provides great value to the users.

As a result most information will be organized after it has been generated, and the shared finding tools will have to work closely with the shared organizing tools. Although most of the organizing will involve people, because machines are not good at deducing meaning, some automated finders may prove useful—for example, by keeping track of visitation patterns and telling you, when you visit a site, what additional sites other people have tended to visit before and after visiting this site. Here the software sidesteps its inability to interpret meaning by giving you information on the search patterns of people who may have the same interests as you and have already done the interpreting.

In the future, finding and organizing tools will handle not just text, but audio and eventually perhaps images and video. These are difficult to classify with current techniques but may yield to new discoveries. You might then hum a tune into your microphone or show a picture to your digital camera and have your tools locate reams of relevant information in the Information Marketplace.

With the amount of information steadily on the rise, the need for organizing and finding tools will increase, and the extent to which they help us find information will affect the future direction and utility of the Information Marketplace.

Computer Security Schemes

Computer security is a broad topic spanning many different issues. However, in practical daily use in the Information Marketplace, computer security will involve three major categories of software tools: privacy, authentication, and payment.

Privacy. You are working on a contract with a client, and you do not want anybody else to see it. But the infrastructure you are using is accessible to many people and computers. How can you ensure the privacy of your communications?

Authentication. All 50,000 employees of your company have just received an e-mail message from the chief executive officer about a sweeping organizational change that will affect everyone. Is the announcement an authentic one from the boss, or is it from an impostor?

Payment. What should you and your bank do to ensure that the checks you send electronically are genuine and to block checks that are not? You frequently rent movies and listen to music over the Information Marketplace; what kind of billing scheme will ensure that the producers of these art forms along with the other middle merchants in the chain get their payment? More generally, how will payments be handled as billions of transactions flow over the world's information infrastructures?

Let's try to address each of these needs with an appropriate software tool. As we do, we must remember that the central problem of computer security lurks in the rich interconnections provided by any information infrastructure, which let people with questionable motives come electronically close to your information. The problem is exacerbated by the digital nature of the information; it is incredibly easy to copy 1s and 0s, and it takes only a few seconds for a "bad" computer to try a few million ways to cross your virtual doorstep. Unscrupulous people are guaranteed to devote substantial computer resources to try to break into your information. If they can do so, they may sell your contract to a nosy competitor, forge your signature, steal some of your money, or snoop into your activities.

Fortunately, there are defenses we can erect to protect ourselves in all these situations. The defenses are derivatives of cryptography—the wartime and spying craft of scrambling or "encrypting" information so that unauthorized users can't make any sense out of it but authorized recipients can unscramble or "decrypt" it in a straightforward way.

In the thousands of years that cryptography has been around, the world has had ample opportunity to develop different schemes and assess their strengths and weaknesses. The simplest codes to construct, and also to break, involve changing every letter of a message to a substitute letter. Changing every letter with the one that follows it in the English alphabet would result in a sentence like this:

zpv hpu ju (you got it)

A computer can decipher even a long message encrypted by letter substitution in only a fraction of a second of computer time, because the machine can exploit the known frequency of letters in English and can keep applying different letter-substitution schemes until it finds one that decodes the text into actual English words (as identified by its dictionary).

Of course, some cryptographic encoding rules are much harder to discover. The toughest codes to unscramble involve changing the substitution rule in a totally random way every time a new block of letters is sent. As long as the recipient has access to the succession of these substitution rules, he or she can decrypt the message. That succession is usually recorded in a "one-time pad" shared by sender and receiver, so named because each page is thrown away after it is used.

Regardless of how hard it may be to break mathematically, any cryptographic scheme is vulnerable to other kinds of attacks. For one thing, perpetrators can infiltrate your camp posing as supporters and compromise the ways in which the codes are made, shared, and disposed, tipping off the enemy. More devastatingly, a person trusted by the cryptographic enterprise can simply be bribed or co-opted. Techies enthralled with novel mathematical techniques for encryption will do well to remember that the same age-old human frailties that have helped win and lose wars carry over unchanged to the modern Information Marketplace.

Most cryptographic schemes lie within the above two extremes of simplicity and mathematical impossibility. With the cost of computation relentlessly dropping, their true cost will increasingly center on ensuring that the schemes are designed to be secure and are well managed. A commonly used standard in the United States, supported by the Department of Commerce, is called DES (for Digital Encryption Standard), developed by the National Security Agency and IBM in the 1950s. It uses a secret cryptographic code, a 56-bit number called a key, which the sender puts into the encryption software module to scramble a document, a picture, or what-have-you. By using the same secret key, the receiver decrypts the scrambled information.

This approach keeps information sufficiently secure from casual eavesdroppers but not from dedicated code breakers. In fact, some people suspect it was constructed in a way that makes it breakable, presumably by U.S. government agencies that wish to monitor the communications of governments that have adopted the code.

Governments may want to break other nation's codes for national security reasons. At the same time and for the same reason, they are anxious to ensure that their own codes are unbreakable. We will revisit this schizophrenic posture and its consequences when we discuss government in chapter 10.

Cryptographic codes like DES work well among a few people but are more problematic in an Information Marketplace that supports millions of users. Keys must be generated and distributed to people who want to communicate. This is usually done by key distribution centers, which naturally become a critical part of the overall security picture, because they must be trusted to keep the keys secret. Furthermore, the total number of keys is huge, because a key must be issued for each *pair* of people or machines that want to communicate. In a global Information Marketplace of one billion computers (our forecast for 2010), the logistics of a system like DES become prohibitively complicated.

Fortunately, in 1976 the scientists Whitman Diffie and Martin Hellman at Stanford University invented a new approach to cryptography that involves two keys instead of one, eliminating these difficulties. Based on this invention, three scientists at the MIT Laboratory for Computer Science—Ronald Rivest, Adi Shamir, and Leonard Adleman—came up with an approach in 1977 called RSA (after their initials). It constructs the pair of keys using the product of two very large prime numbers. Incredibly, the best way to break the code is to solve an impossibly difficult mathematical problem (finding the two huge primes). For small prime numbers (100 digits) it would still take a few hundred years for a hefty computer to crack the code by trial and error. Choosing larger primes makes the code harder to break but also increases the time and cost needed to encrypt and decrypt messages. In this scheme, the users themselves generate the keys without relying on a central authority to do so. And the total number of keys is much smaller, because only two keys are needed for each person—for 1,000 people, all of whom wish to communicate with one another, this amounts to 2,000 RSA keys versus some 1,000,000 DES keys!

There are many encryption schemes around and programs that implement them. They have names like PGP (Pretty Good Privacy), Unix Crypt, Idea, RC4, Ripem. There is even a gadget, the U.S. government's clipper chip, that was supposed to be placed in every cellular telephone, TV set, computer, and other appliance that can communicate. The clipper chip contains the user's key, which would be issued by the government, etched on a solid-state chip. It has been set aside following strong protests by the technical community over its effectiveness.

We'll focus here on the public encryption scheme, because when the dust settles, it or a close cousin of it is the likely leader among several schemes that will meet the security needs of the world's information infrastructures.

The idea behind RSA public cryptography is this: Every person and organization in the Information Marketplace creates two long numbers called keys. One key is "private" and is kept secret by its owner. The other key is "public," which means that it is openly and widely available to everyone, for example, through published directories. Either the private or the public key can be used to encrypt a message. To encrypt using your private key, you will normally type a short password. This causes your machine to use your private key (which is itself securely stored in your machine) to encrypt the message. Once encrypted, a message can only be decrypted with the remaining key. Security middleware in your computer will use these keys to encrypt and decrypt messages, long documents, pictures, and even videos.

Let's see how a system like RSA addresses the three principal issues of Information Marketplace security:

Privacy. To ship that important contract to your client with full privacy, you first encrypt it with your security middleware using your client's public key, which is widely available. Your client then uses the private key in her security middleware to unscramble and read the document. The scrambled message, if intercepted en route, cannot be read, because no one else has your client's private key. The scrambled message cannot be decoded, because then an impossibly difficult mathematical problem would have been solved. Society would then have to decide between punishing as a thief or honoring as a celebrated mathematician a person who somehow cracks the code.

Authentication. To ensure that the ominous memo announcing cutbacks comes from the boss and not from a disgruntled employee, the boss would encrypt his memo using his private key and would send the scrambled message to everybody in the organization. Any recipient would then be able to use the boss's public key to unscramble the scrambled message and therefore be assured that the message could only have come from him. What the boss did was to *digitally sign* the memo. Digital signatures are very important, as we will soon see.

These two approaches can be combined. To send your client a private contract that has also been signed by you, you first apply your digital signature by encrypting it with your private key and then you make it private for your client by further encrypting the resultant scrambled message with your client's public key. Your client will then use first her private key to decode the digitally signed contract and then your public key to authenticate the signature and get the original text.

The computations performed by the security middleware in the sender's and receiver's computers can take several seconds with public cryptography schemes like RSA. To speed things up, you may be tempted to use a faster but less secure scheme like DES. This might be acceptable to you if you could change the key of the less-secure system often enough so that even if an interceptor obtained your key or cracked the code, the violation would be short-lived. You might decide to change the DES key every morning, though that means you would have to send the key to your client. No problem: you use the more secure RSA scheme to send the new DES key to your trusted client.

Before we discuss payment, let's take a look at how the ability to digitally sign information is crucial to the security of public cryptography key distribution schemes. The effectiveness of public cryptography schemes like RSA relies heavily on the two keys that every participant uses. Fortunately, you create your own keys, based on your own arbitrary choice. This eliminates the potential for tampering with the key generation process that characterizes centralized key distribution schemes. Once you have generated a pair of keys for yourself, it is important that you keep your private key from being discovered. If you lose it or suspect that it has been somehow copied, you can always generate a new pair of keys. You must also ensure that in the eyes of the rest of the world your person is clearly associated with the public key that you have generated. Otherwise someone could impersonate you by pretending that his or her public key is yours.

The process that ensures that your public key is indeed yours is called *certification*. It is basically a statement to that effect, digitally signed by a trusted party. This could be an organization specializing in such certification or another person. This is exactly the way we formally convey trust in our society: Someone that we trust, perhaps the government, a notary, a bank, or a friend assures us in person or in writing that some information we wish to have certified is genuine. Birth, marriage, and death certificates are handled that way, as are drivers' licenses, airplane tickets, credit cards, letters of credit, affidavits, and letters of recommendation. The beauty of the public key certification approach is that it carries this familiar method of chaining human trust to the Information Marketplace.

Payment. A primary issue at the heart of the Information Marketplace is the manner in which payments will be made for the purchase of information and information services. One important and lasting way will be similar to what we do today when we sign a check or a credit card. We can't send a picture of our signature, because somebody could steal the bits that represent the picture and forge it. But we can use the encryption technologies to do so through *electronic checks.* Here's one way the scheme would work: Like today's paper checks, a message we generate lists the payee, the amount, our bank and bank account number, the serial number of the check, and the transaction. We digitally sign this check by encrypting it with our private key and then send it to the seller. The seller, using our public key, can unscramble the document and ensure that we have indeed sent in our payment. The seller then forwards the same encrypted document to the bank, which, using our public key, clears the check, charges our account, and credits the seller. Throughout this process anyone can verify, by using our public key, that we are the only one who could have signed the check.

This scheme can become even more secure. For example, the seller can take our "signed" check and encrypt it further with his private key before sending it to his bank, thereby "endorsing" it for deposit to one of the seller's specific accounts. We could even achieve greater security than there is with today's paper checks. In my check to the seller I could encrypt my bank account number with my bank's public key. This way, the seller would be unable to see my bank account number, but my bank would easily see it using the bank's own private key.

This scheme can be extended in a straightforward way to handle credit card transactions and *any* other existing payment methods that require our physical signature. We simply use our digital instead of our physical signature.

Besides these modifications on old-fashioned but established ways of paying, we are likely to see more novel approaches. Perhaps the most radical involves "digital cash," or electronic cash, a new form of tender that can be passed from buyer to seller almost like real cash. You basically pay by using a number, which the electronic cash company gives you when you send it a check. Thereafter, you pay by using this number with merchants, suitably encrypting it so they cannot use it without your authorization. It's in all respects like a check written to a bank except that it can be made untraceable, like real cash. On the other hand, it's not exactly like real cash because a private company rather than the government stands behind it, and you still have to communicate with some central database to use it.

Another approach under development involves "micropayments"—electronic transactions that convey very small amounts of money, in units of one-thousandth of a dollar. Proponents imagine a world where you can rent a song to be played over your computer for 60/1000th of a dollar, or an apple pie recipe for 214/1000th of a dollar. Over time the micropayments would accumulate to the rightful recipients. This scheme is not like checks or digital cash, because it is constructed to involve almost no cost overhead. Whether micropayments become widespread or not will depend greatly on the distribution schemes that will dominate the purchase and sale of information. If, for example, songs and other information goods are sold or rented by brokers, then these intermediaries will aggregate payments and purchases in larger chunks, making micropayments unnecessary.

Another approach involves so-called *smart cards*. These resemble credit cards that have small processors and memories embedded in them. Smart cards can be used in many ways. In France, for example, 22 million such cards were in service in 1996, mostly as credit cards used to authenticate the identity of the purchaser. At the store, you put the card in the slot and, unless the amount is over the limit, the transaction is concluded locally, and rapidly, without calling out to check the card. In other emerging uses, you put a so-called *smart debit card* in your ATM's slot and charge it up with some money, say, $1,000 from your bank account. You then go shopping. When it's time to pay for a $200 item, you insert your card into the store's card reader. It debits your card by $200, reducing the remaining "balance" inside your card to $800. Other information on the card authenticates you and assures the store that the bank will honor the amount that you paid.

Smart cards are likely to emerge for a large number of applications because they speed up transactions—the store need not check with the bank if there is money on the card—and avoid sending vital information about you flowing over the wires. In a futuristic approach now in the research stage at the Siemens company in Germany, you will have a super-smart card that in addition to all its computational capabilities also remembers your fingerprint. When you hold it with your thumb covering the lower-right corner, the card reads your actual fingerprint, compares it to your thumbprint, stored digitally inside the card, and then clears whatever transaction you want to conduct, whether it is withdrawing money from a bank, charging a purchase, or opening the door to your home. If your thumb is not pressing on the right spot, the card is useless. And if the thumbprint does not match the one stored in memory, any attempt to use the card will produce an alarm message warning that the wrong person may be trying to get away with theft. It is even possible to make the card's solid-state fingerprint sensor monitor the blood pulsations within the thumb so that no finger replica can be used by an impostor. The beauty of this scheme is that no one else has access to your precious fingerprint information.

No doubt, we will see a proliferation of both old and new payment mechanisms in the Information Marketplace. Variations will arise from the world's different payment practices; in some countries bank transfers are used almost exclusively in lieu of checks; in others, the entire economy is predicated on cash. In spite of all these options, I have no doubt that the established banking and credit card system, which has been working on computers for over twenty years, after it is suitably modified with digital signatures, will continue as the dominant method of payment. The relatively high current cost to process such a transaction on paper will be trimmed by the very mechanisms of the Information Marketplace, ensuring that this familiar system will retain its dominance as it evolves. The trend is already here—in the late-1990s, a bank's cost to process a simple transaction like a withdrawal or a check the old way from teller to back office is $1.40; for an ATM transaction it is $0.45, and for an electronic bank-to-bank exchange, between $0.02 and $0.08. The established banking and credit card systems will at best be augmented rather than replaced by the new schemes.

Security middleware tools for privacy, authentication, and payment will be actively developed in the coming years and will become an integral part of tomorrow's information infrastructures.

Regardless of which technologies eventually persist, we will surely have a wealth of tools that can satisfactorily address our human needs for security.

Information Infrastructures

Having examined the various information pipes in chapter 2, the human-machine interfaces in chapter 3, and now the basic categories of shared tools, we are ready to assemble in our minds a model of what the information infrastructures of tomorrow will look like. To understand the model is to understand how the Information Marketplace will work.

An information infrastructure has three layers: pipes, tools, and interfaces. So let's imagine a three-story building. We, the users, are standing on the roof along with our applications software.

The bottom floor of the infrastructure contains all the pipes supplied by the world's information carriers—the phone lines, video cables, satellite links, and wireless communications channels, along with the Internet and Web software and protocols that control and utilize these pipes. The top floor of the three-story infrastructure contains all the human-machine interfaces that we on the roof use to get into the building.

The middle floor of the infrastructure is where all the shared middleware software tools will reside—the ones that provide for automatization, e-mail, groupwork and telework, pipe managers, hyper-organizers and finders, and the various types of computer security for privacy, authentication, and payment.

This simple model is easy to visualize. But you won't see it in any one place, because it will be widely distributed. The pipes and the software that controls them will reside underground and overhead and in the computers of the telephone and cable TV exchanges. You will see at most the tail ends of these pipes in the boxes that the carriers will install in your home and office, much as you see today's telephone wall jacks. And you will see the software in your computer that handles interactions with the Web and the Internet. The human machine interfaces will all be close by as gadgets that you can touch and as software inside your computer. The various applications will also be inside your computer, just as they are now, except that they will know about the Internet and Web and software tools, which they will use liberally to carry out their tasks. The shared software tools—the real workhorses of the information infrastructure—will be inside your computer too, widely shared by you and your applications.

This infrastructure model is not one monolithic entity catering equally to everybody's needs. It is really many interconnected infrastructures. A doctor standing on the roof may access the medical infrastructure; its interfaces, tools, and pipes will reside on the three floors but perhaps all along the left side of the building. A vacationer on the roof will use a separate infrastructure; it will use different interfaces, tools, and pipes on the three floors but perhaps along the right side of the building. Usually, these two infrastructures will remain apart. However, because the floors run laterally, different people on the roof will be able to dive down one infrastructure as far as they need and connect to another infrastructure. If a vacationer using the kiosk in Ruby Creek to connect to the travel infrastructure feels ill, he may use the same kiosk interfaces to jump over to the medical infrastructure to get help. Both activities might use e-mail and pipe managers on the second floor, and both will use the same pipes on the first floor, as will the rest of the users up on the roof. This ability to share common resources by running laterally across a floor is made possible by the shared conventions and standards that we have been talking about, which are an essential part of a good infrastructure. Then again, at times people on the roof may simply be busy working on their own computers with standalone applications, unconnected to any infrastructure.

Unlike a real building, our infrastructure building has boundaries between floors that are not rigid. Speech software like Galaxy will operate in part on the third floor as an interface and in part on the second floor as middleware. Most important, as standalone applications become very popular and widely shared, they will become part of the infrastructure and move down to the second floor. For example, a standalone visualization program that we now might use on our own computers on the roof may, if successful and used widely, become middleware and sold widely by software vendors. This migration of new software that will be tried first outside the infrastructure and then move into the infrastructure on its own merits is the free market way through which the infrastructure will evolve from its timid beginnings today to its powerful and more stable forms well into the twenty-first century.

We can now picture how this superb new machinery—the information infrastructures that make up the Information Marketplace—will work. The hundreds of millions of people and their computers on the roof will buy, sell, and freely exchange information and information services using their interfaces and application programs. Many of the "users" will be unmanned computers running automatization procedures on behalf of their owners. The users will employ interfaces and dive into the middle floor, where a host of shared software tools will coordinate their desires, and transport the information required to fulfill those desires through the pipes on the bottom floor. At any given time, these millions of people and machines will be carrying out their designated activities, spanning an equally huge number of economic and personal objectives. The Information Marketplace will be fully active, and its infrastructure will be at work, humming along just like the power, water, gas, telephone, and security infrastructures of a modern city.

It is hard to believe that all of us people on the roof will enter the twenty-first century still isolated from one another by the computing environments, interfaces, and mind-sets of yesterday (DOS, MacOS, OS2, Unix, Windows) that were designed to control standalone computers. No doubt, new systems will emerge that will better exploit the new interconnected infrastructures as incremental improvements of these older systems. We will even see some old systems in new clothes, like the much talked about *network computers,* consisting of a screen and processor but no main storage disk. Users of these devices will store their data and software on big shared machines and, therefore, benefit from lower costs and automatic software updates—so the theory goes. Network computers will prove useful in certain niches, especially where centralized control is acceptable. But they will not challenge full-power PCs, because they are not much cheaper, and more important, because people would rather own than "rent" their resources. Once we got used to cars, it was hard to revert to using buses!

Whether of the network or full-PC variety, the emerging machines continue to be enveloped by computing environments that are only incremental advances over their standalone predecessors. That's not good enough. Until Xerox Parc and Apple came along, all computer users had only one environment in which to interact with their machines: by typing characters. The Macintosh and Windows environment presented us with icons, windows, and pulldown menus, objects on the screen that could be moved around to perform tasks. This new environment represented a radical change that was first resisted and later widely adopted even by the most diehard fans of pure type.

Will such a radical change come about again? Will there be a new environment tailored to the world of distributed computers with novel controls and approaches that will better reflect the new world of information? I firmly believe so. The information infrastructure begs for it. And users are ready, too. In fact our lab and several other labs around the world are doing research on this problem as this book goes to press. The principal characteristic of such an environment is that it would let you view and act on the information in other machines as if it were on your own machine (subject to permission by the other users). This "seamless" integration between local and remote sites would let you share information, assemble data and procedures out of pieces that may be all over the world (updated by other people and organizations), conduct group-work, and more with the same ease you combine files and fire up your word processor today.

With or without a new environment, once there are effective human-machine interfaces, useful tools, and an advanced Web/Internet infastructure of big, clear pipes, the Information Marketplace will become a reality. The exciting question then becomes, What will all the hundreds of millions of us standing on the top of the world's information infrastructure actually *do* with it? How will it change our daily lives? On to part II.

PART II
How Your Life Will Change

5
Daily Life

Music Match

Though the Information Marketplace will dramatically alter the economy and government, and the composition of nations and cultures, each of us still cares first and foremost about what happens to us each morning and evening. People had the same perspective during the Industrial Revolution. They cared little about such abstract matters as the development of modern manufacturing techniques and distribution systems. But they cared deeply about the practical results of these developments: the cars, lights, telephones, and jobs. So before we assess global issues in part III, let us focus the chapters of this part of the book on how the Information Marketplace will transform the way we live, work, and play—how we wake up in the morning, shop, invest our money, entertain ourselves, create art, improve our health, educate our children, perform our work, and join or interact with the organizations that employ us, sell to us, provide community, and expect us to pay taxes.

I'd like to begin with a familiar point. The media are constantly feeding us exciting and imaginative stories about how information technology will change our lives. Most of them are pure fantasies; it's not difficult for people even mildly aware of current technological developments to dream up all sorts of outlandish scenarios. The real challenge is to discern plausible applications that, exciting or not, can reliably develop from today's technological trends, can be economically feasible, and can fulfill some useful human need. That is what I intend to do in this part of the book. I will also touch on some of the more speculative and less likely possibilities, clearly labeling them as such.

To make the visualization of our future lives a bit more fun in this chapter, let's trace the daily activities of an upper-middle-class first-generation American family. Some of the experiences I will describe are only a couple of years away. Others are a decade or more in the future. You will play the part of "father." Your spouse, son, daughter, and parents all figure prominently in your activities. I'll begin each section of this chapter with a scenario from your future life and then explain how the machinery of the Information Marketplace will make the activity a reality, and how soon. As we go, I will also try to highlight the difficulties involved in reaching the more futuristic aspects of these scenarios.

You are gliding over the mountaintops, swaying ever so gently to the faraway sound of an ethereal melody. All is sparkling below. Suddenly something begins to go wrong! The earth dissolves to black. The music seems stronger—right next to you, in fact. The blackness gives way to a dim light, recognizable objects. Yes, you are in your bedroom, it is dawn, and your dream is over. The music is coming from somewhere inside your bed.

Alas, it's time to get up. Your senses have tuned in, you rise, and you find yourself concentrating on the music. It is quite captivating. You smile; you've never heard the piece before, but once again your wakeup service has found an engaging selection. They do this often, yet it still intrigues you. Even when you made a mistake several weeks ago in updating the kinds of music you prefer, they somehow caught it based on your prior two years of preference patterns. You wonder how they took your initial e-form, where you spoke the names of five songs you really like and provided a few other details like your age and home location, and managed to wake you up every morning with melodies that seemed to get better each time.

You are curious about the name of the song you are hearing. As you lean toward the small microphone adjacent to the bed's built-in speakers to ask for it, you wonder if they use your questions as cues to refine your musical profile. However they do it, they are worth the twenty cents a day they charge. You recall that as a child you were thrilled with your stereo jukebox that held three hundred CDs and could wake you up with the songs you chose, but after a while, programming your choices got to be a chore and you got tired of hearing the same tunes over and over. Buying new CDs also cost a lot more than this scheme of piping inexhaustible choices your way. Your service informs you the song title is "Beyond Desire." You'll probably be humming it to yourself all day.

The wakeup call scenario is quite realistic: The bandwidth required to handle music, even at high fidelity, is already available on current telephone, cable, satellite, and wireless carriers. Renting music for special purposes from some Information Marketplace service is also quite possible. There are large computers already available that could each serve several thousand people in this manner.

The profiling of individual tastes has already been established on a smaller scale: you list a few movies you really like, and a program suggests a few more you might enjoy. The entertainment industry could extend this technique to a far more ambitious level. Here's how it could be done: Several "dimensions" of interest, such as jazz, rock, country, and classical would first be defined. Then each genre would be further refined by time period (such as medieval, Renaissance, 1950s, 1980s), type of performance (band, orchestra, quartet, solo), and so on. Some one hundred such features might be used to characterize every piece of music ever recorded. It's a lot of work for the companies offering this service initially, but once it's done and your combination of desirable features has been established, such a system could search through vast musical archives to come up with a new selection each day that would conform to your preferences. Each time, the system would modify your designated features based on your feedback about which selections you like and dislike. Of course, you need not stay shackled to this setup. If you are adventurous, you can always ask for surprises and the system will comply by randomly searching for different, even totally "opposite" profiles.

As we discussed in chapter 3, asking the machine questions in your native language is within reach as long as you stick to the narrow domain of songs and music, which you would. Paying for this service would be straightforward, as would the distribution of royalties to songwriters and artists; there would be no need for the micropayments scheme we discussed, because the royalties paid by the service to artists and music suppliers would accumulate over time to reasonable amounts. The service fee may not be held quite to twenty cents, because all this overhead would have to be amortized in addition to the cost of a five-minute transmission. So maybe it will be fifty cents a day. All in all, a doable proposition. I expect to see it by the year 2000.

House Doctors and Data Sockets

On to the torture of the treadmill. No video, no audio today while you exercise. You want to focus on the running, and you need peace to think through some tough choices ahead. When you are finished, the cheery voice of the "treadmill doctor" congratulates you and tells you that you are on target with your chosen weight maintenance program.

After your shower you wonder what you should wear. It's a real-life meeting day, so you can't work from your home office. In fact, you are expected at headquarters in an hour and a half. You ask your bedroom monitor, built into the top of your dresser, for advice. Fortunately, headquarters provides a "data socket"—an e-form that is always available and that you can use to get up-to-the-minute information—to handle this vain and archaic question. Your bedroom interface contacts the dress information data socket at your company's computer, which reports that the meeting is "Informal. No clients present." The message is posted on your dresser screen, along with still images of you wearing the three suggested outfits it came up with, based on available clean clothes and some simple rules of good taste that you gave it when you first got it. You chuckle, remembering that only last week you asked the same question and the speech-understanding software misunderstood you, thinking you wanted to know the snow conditions at the company parking lot. It went to the wrong data socket and came back with, "Be sure to bring your shovel."

The doctor in the treadmill is easy to configure, because it deals with simple measurements of your weight, pulse rate, running mileage, and speed compared with a desired plan. The software you got with your treadmill lets you script the automatization procedure that fires up every time you step onto the machine.

The attire and weather data sockets are nothing more than permanent e-forms for employees' automatization procedures. Each consists of a list of fixed entries like "meeting time," "meeting dress code," or "parking lot conditions" and their variable values like "9:00 A.M.," "formal," "informal, no clients," "no snow," "heavy uncleared snow."

Any self-respecting teenage programmer could whip up a data socket like this. Yet there is incredible power in them that comes from the agreements that they signify—everyone working for this company, and all of their computers, can count on these sockets being there and being properly updated every day. They can thus "plug" their customized automatization procedures (whatever they may be) into these sockets and use the socket data in whatever way they like for their purposes. Because the data socket always uses the same words for the same shared concepts, your dress advisor program can be made to understand the meaning of "informal no clients" by being programmed to automatically search your clean-clothes inventory for sportshirts and slacks that match the designated level of dress and satisfy your basic rules, like no plaids on pants or shirt. At the homes of other company employees, automatization programs, scripted by them, take actions that suit *their* personalities and purposes.

Simple and free of pizzazz as these data sockets are, they nevertheless offer a lot of utility. A good deal of information technology is or can be like that. So, techies and humies alike, don't be afraid of appearing "simplistic" or behind the times. If you can find one of these very simple, very useful applications of technology to human needs, you have hit on a real winner. Go for it!

Both the treadmill doctor and the data sockets can be built today. The difficulty with the data sockets is that they require human agreement and decisions as to what information should be selected for these shared communications and what value it would add for the participants—barriers that have been consistently more difficult to overcome than technological ones. Nevertheless, that too will happen but could take a decade.

Before we leave this section, let me go back to the point I made earlier about exciting but unrealistic scenarios. It's quite easy to add to the above scenario this intriguing possibility:

> Fully awake, you head for the bathroom. You brush your teeth, and some of your mellow disposition starts to disappear. That terrible sink is at it again. It has detected minor traces of blood from your gums and is now scolding you in a deep parental voice: "At the rate you are going there is a fifty-fifty chance that you will have a periodontal incident in twelve to fifteen months and a loss of half your teeth by the time you are fifty-five years old." Mumbling to yourself, you reach for the rubber tip and hope for the best.

The sink "doctor" is much tougher to construct than the treadmill doctor, because it is difficult to reliably and automatically analyze an unknown amount of liquid, especially with all the uncontrolled foodstuffs and chemical substances that it might carry down the drain. It is not possible today. If it ever comes about, it may be more expensive than you want.

Auto-Cook

> Divine breakfast smells fill your nostrils. Your spouse, JoAnne, has just gotten out of bed. Oh yes, she said last night that she would be taking the day off from her own home office work in order to do some extensive

shopping. You join her in the kitchen. It seems quiet this morning because your twelve-year-old son, Nicholas, has stayed overnight at his best friend's house and will go to school from there. The auto-cook is out of order again, and JoAnne mumbles her favorite aphorisms about machines as she fries the eggs manually. Fortunately, the local egg shortage has not affected your household since your food-supplies monitor found eggs within your maximum tolerable price in a nearby state, bought them, and had them delivered to your house's produce bin last night.

The Information Marketplace has little to do with automatic cooking. That's the job for a robot that can cook a meal given a recipe and the right supplies. But the Information Marketplace plays a role in giving us the ability to tell the auto-cook and the many other auto-appliances in our homes what to do and to let them talk to each other directly. So let's see how it might be done.

To prepare your meals, the auto-cook must be able to select specific foods. This is easiest if done from specially fitted bins of your house. It must then cut, mix, and process these supplies—a bit trickier, because some foods are difficult to handle, but not impossible. The unit must then move the food into the right receptacles, heat them, perhaps stir, and so on. All this, again, is plausible.

If we think of a robot that looks like us and does what we do in the kitchen, we are thinking in the wrong direction. Mimicking human movements, while appealing to our imagination and to our research aspirations, is still too complex mechanically and computationally. What is easier and more likely to work is a collection of dedicated and far simpler robots that are part of the microwave, stove, and sink and have bearings and levers that can manipulate foods, pans, and utensils. Such robots are just an advance of the same approach that is used in clothes washers and dryers: Mechanical hands don't scrub and squeeze our clothes and pin them on clotheslines. Instead, motors, agitators, squirters, blowers, and revolving cages perform the washing and drying functions. Ditto in the kitchen.

The food monitor that found eggs is technically feasible today. It is nothing more than an automatization procedure that auto-dials nearby supermarkets, looking via their e-forms and data sockets for the desired supplies and closing a deal when it finds what it needs in a price range you told it was acceptable.

Food supplies can be delivered by a special supermarket delivery vehicle that unlocks your house's food supply bins and delivers from the outside all the food supplies that your kitchen food inventory program has indicated are low. Deliveries of ordinary supermarket food bags that are ordered by e-mail or phone are already under way in several experiments in the United States. In 1996 they typically cost around seven dollars per delivery plus 5 percent of the total bill. Technically, this arrangement is straightforward; it requires mostly human agreements, minimal software, and a few electromechanical gadgets rather than new technology. Detecting low inventory status, for example, can be done by ancient weight-activated switches in each bin. Communicating this information to the supermarket, and to your bank to process payment, is done by simple automatization programs, which can also communicate any special orders, say, for extra cheese, grapes, and salsa in the event of a forthcoming party.

The calculations for the quantity and amount of food to buy and prepare would be driven by your standing procedures—verbal instructions about how many people to feed ordinarily. Corrective mechanisms would allow you to tailor future procedures and recipes to your tastes if you want the soufflé to have a crispier top or the potatoes a more liquefied sauce. The auto-cook would remember and oblige.

The best way to implement all this is to build a new kind of kitchen with new kinds of ranges and sinks full of automated devices. This approach simplifies the robotics difficulties, though it may place extra constraints on kitchen space and functional design. It shouldn't, but it might also make it more difficult for us humans to simply grab a pan and fry an egg. Great cooks will probably shun this machinery, but many of the rest of us will delight in the convenience of having food ready when we arrive home. The automated kitchen will surely come, much as those automatic bread-making machines have suddenly appeared before us. I am surprised it hasn't happened yet.

The auto-cook approach can be extended to another everyday task: cleaning. Again, a human-looking robot is not yet technically feasible or affordable. If we built it with current technology, it would work only part of the time and would also trip over furniture and break things. There are more effective approaches. My best friend, Linos, who hates to do laundry and loves cool sheets, invented forty years ago the continuous sheet—it goes around your bed and down below it to a dedicated cylindrical shaped washer/dryer and returns up the other side fresh and ready to re-cover your bed. As you toss and turn, the ratchet mechanism feeds a new cool sheet under you. In the same spirit, special vents where the floors join the walls could be built to remove dust through suction or to emit cleaning solutions for the bathroom and kitchen floors. And the interiors of walls could be equipped with pipes that enable small robots to go where they are needed without tripping over things along the way. Again, these developments rely mostly on electromechanical innovation; the Information Marketplace may well play its communications role long before the robotics become reliable and affordable.

Your Own Custom Publisher

Breakfast is almost over when the large video screen crystallizes on one kitchen wall and you see up in the corner that a news briefing at this time will take from two to six and a half minutes, depending on the level of detail that you wish to pursue. You want to get going, so you ask for the condensed version. The combined BBC-CNN summary on global peace initiatives follows. You are then heartened by an affectionate tidbit about your schoolmate John's engagement to Maria, complete with pictures, which your system "mined" out of your tiny hometown's electronic *Sentinel.* You ask for an additional topic or two, as does your spouse, who wants to know a bit more about that killer sale at Deere Tractor. Finally, you head for your car.

The huge video screen, often called a window-wall, is something techies must make a lot more progress on, at least if you want it to be thin and affordable. For now, the only way to present you with a very large high-definition video display is to piece together a monster made up of many ordinary screens that weigh a ton or a projection system that would cost more than the room and need more than the room's length to fill the wall with its images. Compelling, bright, sharp, and attractive window-walls will eventually come, if in no other way than by stitching together liquid-crystal display panels like those on laptop computers. One exciting and promising new display technology, pioneered by Texas Instruments, involves a solid-state chip the size of a postage stamp on which are etched as many as two million tiny mirrors. The mirrors are attached across their diagonals to equally tiny electromechanically driven torsion bars. Under computer control, some of these mirrors are moved, deflecting the bright light that hits them and forming a reflected image on the wall through appropriate optical lenses.

Tailoring news reports to your interests by computer is old news. In the mid–1980s, Dave Gifford of our laboratory, co-founder of the company Open Market, built a working system that he called the Boston Community Information System. It consisted of several hundred palm-sized FM radios and an equal number of users who attached these radios to the backs of their PCs. An FM station was commandeered, and its broadcast signal was doctored to transmit text along with its normal broadcast. The text transmission did not interfere with the audio picked up on an ordinary radio, but it could be detected by the special PC receivers. Gifford then obtained a continuous feed of various news wire services like the *New York Times* and the *Associated Press*. He provided an ambitious piece of automatization software to each user so they could tailor the news to their individual wishes. If someone was interested in sports and Man-hattan, they would type these into the program. Overnight, the computer would select from the barrage of wire stories only the ones mentioning sports and Manhattan. In the morning each person could read his or her screen or print the tailored news and take it to the breakfast table. Gifford's system worked so well that many people begged to join. Eventually the idea went commercial; the Lotus company used it to update users' data with custom stock quotes and news. Of course, when the successors to Gifford's systems are offered in the Information Marketplace, you'll be able to get text or audio or video, depending on what's available, what you need, and how much you want to pay. Furthermore, these new systems should be able to monitor many more news feeds, including local sources of your choice, wherever they may be.

One Sunday the Boston Community Information System earned its keep in a half-funny, half-serious way. I was preparing to leave for Europe when I got a frantic call from Gifford. His machine, which was tailored to catch news about computers and our lab, among other things, had alerted him to a story that had just come on the wire. It previewed another story that was still embargoed but would allegedly be on the front page of the *New York Times* the next morning. The story would say some terrible things about MIT's "Computer Lab" and its activities in Japan.

Panicked, I called the weekend number of the *Times* and asked to speak to whoever had written this patently false story, because we had no activities in Japan at that time! I was quickly dispatched to the editor on call, who was flabbergasted that someone out of the blue would call about a story that was supposed to have been kept secret. He begged me to tell him how I had found out about the piece, and I begged him to tell me what the story was about. We made a deal to exchange information. Hearing about the news wire attributions, he concluded that the story had accidentally been leaked to the wire services by a young journalist. I was relieved to find out that the story was not about the Laboratory for Computer Science but another MIT lab; neither was called the "Computer Lab," yet both dealt with computers, hence the confusion of the filtering program. I promptly put out an urgent trace for the other lab's director, found him half a continent away, woke him up, and alerted him to the ugly story. He instantly recognized what it was about and said he would call someone he knew at the *Times* to stop this stupidity.

Having done my good deed, I was about to leave for the airport when Gifford called back. His system had just alerted him of another one-liner on the news wire saying that the earlier message was wrong and that no story about MIT's computer lab would be printed on Monday. I couldn't resist the temptation—I called the reporter back at the *Times* to tell him that I was glad he had pulled the story. If he had been in the same room, I think he would have attacked me; he screamed, *"How in the world can you know so fast what I am doing at my desk?! I just did this!"* An altered story about the other lab came out a day later, on Tuesday, presumably after some more checks of its veracity by the *Times*.

Informed Automobiles

During your drive to work you get an emergency briefing from your company's computers. It seems that four hours earlier a Chinese official decided quite suddenly to sign a major contract with your company. As a result, it was decided that your morning meeting would be moved to a nearby company site where this new work would be carried out so that you could include it in your plans. You have visited the site only twice in the last nine years, so you ask your car system for directions. It talks you through the various expressway exits and local turns with uncanny timing, giving you the next instruction exactly when you need to know it. The system continues the briefing between driving instructions, providing background on the Chinese official and a few other tidbits that have accumulated during the night in the company's offices around the world.

The company server can communicate with your car and its computer through the communication technologies that we have discussed. Today, computer-to-computer communication using the cellular phone system is spotty, but things will soon be better as the pipe wars sort themselves out. Having your car computer know your exact location is not very difficult. A global positioning satellite (GPS) system receiver mounted on the roof of your car can inform the computer of your exact latitude and longitude.

This works as long as the GPS receiver can see the satellites—it fails in a tunnel or in a city like New York where not much sky is visible from the streets below. Never mind. An old-fashioned dead-reckoning system similar to the ones mariners have used for centuries can come to the rescue. The dead-reckoner counts the turns of your wheels to figure how far you have gone and records the position of your steering wheel to know which ways you have turned. Because your car computer has a map of the area, it is a simple matter for it to take the last GPS fix and then use the dead-reckoner information to figure out where you have gone since. Because errors could accumulate, as on any dead-reckoning navigational system, as soon as the sky reappears the GPS takes another accurate fix, corrects the position of your car on its map, and resets the dead-reckoner. Knowing where you are, where you should go, and the map's indication of one-way streets and other information useful to you and your company, the computer can figure out the shortest route; this much can be done today by the Galaxy system. The system can then issue the correct instructions to you a bit ahead of each upcoming turn so that you have time to react. A simple speech output interface converts these instructions to voice commands.

The reveries on how the Information Age might affect transportation can go on for a long time. There is much talk today about smart highways—roadways that not only bathe your car in information, but take over control of your car while you read, rest, or play games inside. ARPA has already constructed an *autonomous vehicle* that can zoom along a highway lane at sixty miles an hour using its own automatic vision system to detect where the road is headed and to avoid obstacles. The system built by Carnegie Mellon University researchers was used to navigate a minivan from Washington, D.C., to San Diego; 98 percent of the trip was made under computer control, with the longest uninterrupted stretch being a few hundred miles in Kansas.

Another public transportation dream involves thousands of two-passenger vehicles, perhaps electric for environmental reasons, that are scattered throughout a city, plugged into public recharging ports. You'd leave your house, grab the nearest car, drive it where you want to go, and leave it at your destination plugged in again for the next driver, who will surely come along in a few minutes or hours. Because people generally start and end their day at home, the cars may dance around the city, but their overall distribution will not change much from day to day. Variants of this scheme have you calling for a public transportation minicar, which automatically zips to your front door like an obedient dog called from its nearest resting place. You tell it where you want to go, and it carries you there.

Such systems are not integral to the Information Marketplace, and they won't be feasible for many years for technical and social reasons. The egalitarian mass transit schemes are not very likely because of their radical economic, maintenance, and ownership arrangements. The automatic driving schemes are still far from being safe. Moreover, a modest-sized city would have to spend billions of dollars to redesign, construct, and instrument roadways so that such vehicles could operate.

More likely in the next decade are advanced versions of today's privately owned cars, fully instrumented with computer, communications, and navigation equipment. They will have additional novelties like advanced cruise controls that keep your car at a safe distance from the curb and cars around you, alerting you to potential accidents. These vehicles will be full participants in the world's Information Marketplace, enabling you to do much of what you would do from your home or office while you are on the road.

Discovering Insurmountable Needs

As you drive to your meeting, your spouse, JoAnne, is in the living room browsing through L. L. Bean in Maine, Harrods in London, Saks Fifth Avenue in New York, Le Printemps in Paris, C&A in Germany, Mikimoto in Tokyo, and GUM in Moscow. If you were there, you might be choosing different department stores, virtually walking up and down the aisles of Home Depot, Radio Shack, the Hong Kong Super-Mall, or the Ulmia woodworking tool factory in Ulm, Germany. But you are grateful that she is doing the shopping.

JoAnne's thoughts are on your common joke: the discovery of "insurmountable" needs. The rush of shopping somehow seems highest when, while browsing, you find an item that fulfills a dire need that you didn't know you had until you went shopping. Her online gift shopping on the living room computer is almost done when she decides to take one last look at the Deere Tractor sale and, bang, a great need is unearthed: a small lawn tractor with an attractive price. What really caught her eye was the backhoe attachment. Despite her extensive mechanical engineering experience, she has never seen a backhoe so small. She quickly paces through a short video of the machine at work and, discounting for the exaggerated advertising, is still surprised to see the heavy landscaping the compact machine can handle. She asks for a drawing of the tractor, which discloses the extensive use of carbon composite materials in its construction.

With mounting interest, she fires off two requests—one for consumer reports on the product, the other a call for anyone who might be interested in selling her a used model. Two minutes later the responses start trickling in. The consumer reports are quite positive, and the prices asked by the nineteen sellers scattered across the country are high enough to convince her that the machine is really good. Just to be safe, because the money is considerable, she decides to go for a list of actual user complaints. That costs her a few extra dollars, paid to one of the many services that specialize in compiling customer reports on outdoor equipment, but it is money wisely spent. She is happy to find out that the complaints are few and minor. Back at the Deere Tractor site, she asks a few questions about maintenance costs and ease of changing attachments and finds herself controlling an interactive video that deals with these topics.

All right. She takes a deep breath and says the magic words: "I'll buy it." She tells the computer which charge account to use, and it forwards the necessary codes and security protocols. A few seconds later she chooses the cheap and slow delivery option from UPS despite her adrenal desire to go for the shameless option of same-day delivery, also from UPS but at a much greater cost.

The tractor is JoAnne's big birthday present to you, though you're not quite as interested in gardening as she is. Nonetheless, she knows you will be happy, for you have lamented often enough about the need for some serious landscaping around the porch, garage, and garden.

JoAnne is about to leave the transaction when a message appears, courtesy of an intercompany marketing module programmed by the Home and Garden Suppliers Network. It informs her about a snazzy piece of software that purportedly helps in designing landscapes that can be shaped using, among other tools, the very machine she just bought. She asks for an interactive demonstration of the software and puts it through its paces right there on her screen. The software also offers a permanent link to online newsletters and update notices from Deere and all participating companies. The cost is minor, so she commits this sale too. No need for shipping this time; the software is downloaded electronically along with a handsome color manual, both automatically installed on her computer.

Most Americans are inclined to believe that shopping will evolve in this direction, because it seems like a natural extension of shopping by catalog. Some Europeans argue differently. They believe the citizens of the older continent will prefer to shop in person so they can rub shoulders with fellow shoppers and pick up and feel the merchandise. No doubt, there are people who will reject shopping remotely—and not just in Europe. Yet we must not forget that similar assertions were made about some of the biggest retail trends around the world such as fast-food restaurants. Blinded by their surface tribal customs, people have a way of forgetting how similar we all are in our wants and needs.

Several parts of the shopping scenario are already here. Consumer reports are already available on the Internet from several providers. Downloading software is commonly done on the Internet, though it is still slow. However, the swiftness of JoAnne's transactions and the quality of her various ways of shopping are well ahead of the state of the art. In a bizarre conspiracy of bureaucracy, market research, paranoia, and billing jitters, today's electronic shoppers are enslaved into filling out countless electronic forms requiring completely repetitious information and are asked stupid questions like, Are you really sure you want this? Time out for a short tantrum.

The mentality that led to this kind of query is rooted in two evils—the sellers' desire to get away with cheap programming and their desire to keep users from accidentally burning themselves. These annoyances dampen people's willingness to actually complete a purchase online. Such lazy programming practices, and the inane notices, had better be abolished, or sellers won't find any real buyers, just curiosity seekers with time to kill. All that is needed is a bit of real programming with the user's needs in mind and the acceptance of an automatic, secure, traceable burst of information from the shopper's auto-profile that provides name, address, and credit information, complete with the requisite authentications. The buyer should only have to say, "I'll buy it."

The other shortcoming of today's electronic shops is their bland, typewritten description of products or, at best, their awkward still-life graphics. And the ones who delve into fancier visuals are penalized by the delays incurred in sending all this data and building up the images. In time the boring hierarchical menus and sluggish graphics will yield to instantly accessible and attractive corridors and alleys that we will feel compelled to explore . . . always in search of insurmountable needs! The simple text descriptions will yield to multimodal presentations and simulated trials like those we encountered at the World Shop. That kind of performance, including video, would be just barely possible today at a dedicated location like a World Shop, outfitted with fast computers and fast communications pipes. Progress in this quest will be exclusively determined by progress in getting higher-bandwidth pipes to our homes. It will probably be a decade before enough people have access to these capabilities. The important observation, though, is that there are no impenetrable technical barriers along the way, and there are enough people around who will like to shop this way. So get ready for it. It will happen.

Mass Individualization

The above vignette predicts another important aspect of shopping in the Information Marketplace: "reverse advertising," the opposite of old-fashioned direct advertising. You state your requirements—for example, that you want a certain kind of tractor—and only those vendors that can match them come forward with their offerings. Though we are familiar with this practice for used items—that's what the "want" ads are in newspapers—we do not yet use it for new goods and services. Instead of wandering from store to store looking for the right pair of shoes, you publish a precise description of what you are after and make the sellers come to you. Reverse advertising is desirable not only to buyers but also to sellers, because it allows them to focus their energies on the more promising leads. Both buyers and sellers become more productive in their quests. To make reverse advertising possible, special markets will be established within the Information Marketplace where buyers and sellers can match needs with availability. That is exactly what JoAnne did when she posted a request in a market (really, a computer site) designated for used gardening equipment.

Reverse advertising matches existing products and services to emerging needs. A more aggressive approach would involve manufacturers' making new products and generating new services on the spot, custom-fitted to the consumers who order them. This could be done with apparel, furniture, and cars, not to mention financial and legal services. The possibilities and combinations are endless.

JoAnne, feeling the need to get out of the house, drives down to the nearby Auto-Shoe Shop. She would have gone to the World Shop, but luckily her town has an Auto-Shoe outlet. She enters one of the store's cubicles and places her right foot on a special platform. Twenty cloth straps wrap around it. She turns a knob to tighten them to what she deems a desirable level of comfort. She then pushes the proverbial button, and the measurements of her foot are recorded for future use. She does the same for her left foot, which is a bit larger. This will be the only time she has to undergo this procedure, because from now on she'll have the forty standard measurements, which have been agreed upon by the shoe manufacturer's world guild, in her personal shopping profile. With the digital profiles of her feet recorded, she now shops through the store's online shoe catalog. Sure, she could do this from home, but half the reason she came down here was to see some real people. The other half was to engage in the instant gratification of wearing her new shoes out of the store.

A specially fitted screen at the bottom of a mirror shows her wearing each of the shoes she contemplates buying. She chooses a pair, makes a few adjustments like adding brass grommets, and pushes the Make button. She then walks along the manufacturing gallery to watch them being made. The flat pieces are cut by computer-driven cutters (minimizing waste in the process). Other machines glue and stitch the uppers to the soles. A human operator intervenes for the intricate job of affixing a buckle to the uppers. Fifteen minutes later, her shoes are ready. She puts them on and marvels at the fit.

This scenario has not quite happened yet. I first wrote about it back in 1976 in *The Computer Age: A Twenty-Year View,* a book by me and MIT's provost, Joel Moses, with contributions from several authors. In my own contribution on manufacturing, I called the scheme Mass Individualized Production. It creates a strange time warp, combining the Old World custom of hand-crafted products and services with the cost benefits of mass production. Costs were high, and computer-robotic techniques were still undeveloped, preventing this kind of shoemaking from happening in the last twenty years. However, since then quite a bit of progress has taken place, resulting in comparable schemes that have been successful in the making of dresses, suits, and pants. This kind of manufacturing is now gaining popularity under the name *agile manufacturing*. My confidence that mass individualized products will appear stems from several quarters. In manufacturing there is a clear movement toward low-cost customized goods for smaller and smaller niches of users. Online shopping will make ordering easier. And humans have an ever-present need for individual attention that is reinforced by the fact that no two human bodies or minds are exactly alike. As costs inevitably decrease further, and as the technologies mature, mass individualized production will come of age, with apparel being the likely first market.

Another important area of individualized manufacturing involves shelter. Today approximately half the cost of building a house goes toward materials and half toward labor. If the labor component could be reduced, the overall savings would be significant, so much so that it would alter the balance of who can afford to buy a house. Individualized construction would also render living quarters that are more suitable to their owners.

Unfortunately, architecture has remained close to its ancient origins. Though drawing has been automated, houses are still constructed primarily of many, many pieces comprising small units of wood, bricks, concrete, and glass. Building contractors spend a great deal of time finding and fashioning the pieces together to make walls, ceilings, doorways, and windows; threading pipes and wires through the unfinished walls; and finishing the surfaces of everything with hand labor. Suppose instead that walls, ceilings, and floors were pre-built in modules with optional but standard ducts for water, electricity, phone and computer wires, and the like. Suppose too that these large modules could be assembled easily and rapidly, so that a complete house could go up in finished form in a week rather than four months. Right now the costs to prepare the modules are excessively high. There is also resistance from the unions, fearing reduced employment in construction, and from the inertia of ancient building codes that are still in effect. But it's hard to believe that this approach works for making far more complex artifacts like electronic systems and somehow does not work for houses. It's far more likely that people are simply reluctant to agree (or to risk agreement) on this mode of construction and on a set of basic modules so as to step up production and drop costs.

Despite all the obstacles, prefabricated houses are moving gradually from their standard, one-box-fits-all designs to customizable packages. Information technology can be used in the construction process to help design, visualize, schedule, order materials, and perhaps someday drive the fabrication of major house modules with partially automated construction techniques. The wilder innovations of constructor robots and modularized walls, floors, and roofs are probably in the future. Much of the rest is already employed in the construction process. The Information Marketplace will come into play mostly in the front end of planning the construction and will be an aid throughout the construction process.

Once a house is up (with its pre-inserted wiring), it can be endowed with the requisite information systems so it in effect becomes its own mini–Information Marketplace. Besides the auto-cook and automated cleaning gadgets, it would offer a host of sensors and controls for lights, temperature, door latches, burglar and fire and gas alarms; entertainment, health monitoring, babysitting, and package reception systems; live or changeable visual displays (some in lieu of static artwork); multiple communications pipes to cable, telephone, wireless, and satellite links; and outdoor amenities like pool water purification and garden irrigation. Once the electronics are integrated into the house and with one another, you will not notice them any more or less than you notice your present hot water heater, furnace, refrigerator, washer and dryer, and other electromechanical gadgets, The difference, of course, is that the new devices will communicate with one another about their goals and their problems.

You can envision it yourself, now.

You're working late at headquarters when you notify your house computer that unexpectedly you will be bringing over twelve associates in ten minutes for a casual after-dinner meeting. The auto-cook quickly starts heating water for one dozen coffees and teas. The sprinkler system overrules the standard program and shuts down immediately. The air conditioner goes on. And the outside lights go on standby to come up in five minutes, to show your home in full splendor.

Togetherness

It is now afternoon. Nicholas is home from school and sprawled out on the living room floor, busily interacting with his new pal Mbugua in Kenya. They are cooperating on essays for their respective classes; Mbugua's is on Boston and Nicholas's is on Nairobi. Nicholas is having trouble answering Mbugua's questions about the age of Boston compared with the age of Nairobi; his queries on the Net give zero hits on the comparison. He exhales the ageless homework sigh; he must do it the long way by looking up each city on the online encyclopedia and comparing their ages himself. Why don't teachers tell you exactly how to do your work? But the interaction gets considerably more animated and the joyous cries get louder as the youngsters exchange information about the Boston Red Sox and the lion sanctuary on the Tana River. Nicholas looks up to his mother, "Please, Mom, can I go on a canoe expedition down the Tana with Mbugua when school is out?" "That's a very long way off for a real-life trek," she answers. "Maybe you can VR it? I heard there is a good series out on African waterways." You enter the front door, home from your meeting, and close ranks with your spouse. Nicholas will be able to take a VR (virtual reality) excursion on the up-coming weekend; a few queries later, you all find out that the software for the Tana River is available but can't be downloaded so it has to come the slow way. Fortunately it includes real pebbles from the river. Nicholas lights up in anticipation.

You hurried home because you all wanted to have an early dinner. Something special was planned for the evening. Dinner is served and Nicholas reluctantly tells the kitchen screen showing a baseball video to go to "dinner mode." The image disappears. Nicholas has known this rule all his life: no info-port activity while eating dinner with the family. He had once asked you why this was the family's practice, and you re-sponded with your Greek grandmother's favorite saying: "Too much 'Kyrie Eleison' is boring even to God." Nicholas knew that "Kyrie Eleison" meant "Lord, have mercy" but he still doesn't understand why he has to turn off the fun stuff. Adults are so stupid.

Dinner ends by 5:30 so that the family can assemble in the living room for a get-together with daughter Mary, who is studying architec-tural design in Milan, and your parents, who are taking in the warm weather at their vacation spot in Crete. Sometimes it's stressful for your bicultural family to have three members in the States and three around the Med, so you try to spend some time together twice a week. Even though information transport has become less expensive around the world, the computer-aided reunions are still costly—but worth it. It's al-ready very late in southern Europe, so you make the connections. Three big screens in three distant places in the world fill with the faces of your family, linked by satellite and the bonds of love . . . just as it is happening in countless other houses around the globe.

JoAnne is anxious about the well-being of her daughter and your par-ents. After reporting their respective conditions and engaging in some chit-chat, everyone asks to see Mary's new design. She had refused to show it to them until it was finished. Now she leans over to her photo archive and tells it to send the folio of different perspectives to the assembled group. You all admire the graceful lines of her creation, a theater, which has been rendered so faithfully by your respective computers that it looks absolutely real. Nicholas is bored and says so; his older sister is at it again telling everybody what to do. But even he marvels at the majestic atrium with the five-meter fountain ejecting swirling streams of water. The stunning illumination elicits all sorts of praise. Mary excitedly recounts how a famous architect from his office in Tokyo took interest in her work and commented on her progress along the way, making suggestions here and there.

It's time for Nicholas to take center stage. He speaks of his Kenyan friend and the prospective river adventure, to the delight of the grandparents. Your father is intrigued but quiet. He can sense that, like most youngsters, Nicholas and Mary are conscious that their expected role is to please parents and grandparents, who seem to derive so much happiness from these encounters. He knows, too, but is not about to tell them, that after the old ones have gone for good, the young ones will realize with considerable surprise that they themselves were the ones who needed these encounters the most, not their grandparents. Nicholas disrupts the thought: "Grandpa, can you buy me a summer excursion down the Tana River?" You and JoAnne intervene instantly, and Mary is a close second, hampered only by the quarter-second satellite transmission delay. She complains that with all these frivolous requests for money her own legitimate needs for funds will not be taken seriously. After some questioning, you find that Mary had wanted to ask you for—what else—more money. After you find out how she intends to use it, you vow to send the required electronic transfusion overnight.

Children from across the globe already come together for educational purposes on the World Wide Web with text, limited audio, and very limited images. The video component, as for so many other applications of the Information Marketplace, will again have to wait for faster and more affordable transport. Experienced architects are already commenting on the designs of aspiring students. One such experiment is taking place at the MIT School of Architecture under the leadership of its dean, my colleague Bill Mitchell. The senior architects, Frank O. Gehry in Santa Monica, Ove Arup in New York, and their associates, who are working architects, not academics, enjoy imparting their experience to students in a hands-on way. The students gain valuable experience they could never get before, unless they were rich enough, lucky enough, and idle enough to move around the world and visit the studios and creations of their architectural heroes. This kind of collaborative design will surely expand. Already, related work is going on at Kumamoto University, at the University of British Columbia, and at Xerox Parc. We will say a great deal more about educational uses of the Information Marketplace in chapter 8. For now, let's be assured that both Nicholas's and Mary's activities can be supported.

Nicholas's encounter with Mbugua and the family reunion would be carried out by cooperative groupwork modules running on the computers at the various sites. Again, except for the video, prototypes of such encounters have been running for some time on the Internet. A Web application called Mbone already combines audio and rudimentary video today. The video is sufficiently jerky and grainy to justify its status as pioneer, but clearly this type of interaction will come.

Mary's digital archive is pretty much available today; we discuss it in the next chapter. The rendering and blending that she does are also quite feasible. Professor Julie Dorsey, who works in our lab and MIT's architecture department, has developed a complex rendering and illumination program that can portray different views of architectural creations based on different illuminations. She even portrays visually the ways sounds propagate and get reflected in a building.

Here is a part of this scenario that is far more difficult to achieve:

Mary invites everyone to put on their VR glasses and enter the theater through the main entrance. They see the theater as if they were walking into the real thing, as the images move to show this new view.

For this to happen, for Mary's family to be able to actually walk into the atrium under their own control and see on their computer screens what they would have seen in the real building, will require systems that are still well into the future. That's because the computer processing needed to sustain the simulation must operate hundreds of thousands of times faster than today's computers can manage. This improvement will be achieved in part with more clever ways of representing and producing the scene, for example, by a technique called *image rendering* that stores images of different views and stitches them together with computed images as the viewer walks through the space.

The rest of the improvement will be achieved with an increase in the raw speed of computation. Before Moore's Law (explained in the appendix) is invalidated in ten years or so by physical limitations on how small computer circuits can be, it will ensure that microprocessors will become 50 times faster. An additional speedup of 100 to 1,000 times can come from harnessing many microprocessors together, much like using many horses to pull a heavier load. We know how to do this today, so it too will come about, although initially it will be expensive and therefore limited to institutions that do design, like Mary's school.

Computationally intense simulations, such as full-immersion virtual reality with haptic suits, goggles, and trackable helmets, for walking through the building or going down the Tana River, will take one to two decades before they are possible at a moderate quality level and more time than that before they're affordable. So Nicholas's virtual excursion will be crude and will rely a lot on stitched video sequences—still great for the imagination. A breakthrough toward these exciting VR systems is always possible, but there is no direct line yet to a low-cost product, as there is for speech understanding, for example.

Financial Planning

The family get-together ends. Nicholas goes to his room to finish his homework; if he gets it done soon, he'll have an hour to lounge before bedtime. His mother steps outside into the fading evening light to envision what she will attack first with "your" tractor. You take a seat in your office, a separate room attached to the house, and you proceed to access the personal finance software in your computer and transfer the funds to Mary's Milan account. You request an updated financial plan that includes all the exchanges of the day. It advises you that you will be in a cash-flow bind by the end of the month and recommends that you cash in your Alpha company stock. You let out a sigh similar to the one Nicholas emitted earlier; you've been meaning to conduct a few transactions that will improve your financial picture. You can't procrastinate any longer.

You overrule the Alpha suggestion because of the stock's sentimental value. After all, it's your distant cousin's company, and family is family. You ask for the next-best advice and take it, ordering that a small bank

CD be cashed. The program reminds you that although Nicholas's college fund looks good, your retirement fund is still short of target by 35 percent at the current best projection.

You request suggestions, and the program advises you about the initial public offerings of two new stocks that match your investment interests profile, although they represent a higher potential return in exchange for a greater-than-acceptable risk. You access more information from your brokerage, where you can hear recorded advice from the company's experts on each of the stocks as you see charts of performance and forecasts and video clips of the company's products. You know you need to build up your reserves, and because you and your spouse are both still in your early forties, you feel you can take the risk. You check with JoAnne and commit to the deal. Your stock management module looks around for the best fees for this exchange and reports that they are a full 25 percent below what your broker wants. You promptly redirect this information to your broker's office, where an automated procedure responds that they too will honor a 25 percent discount. You ask for 30 percent, but the program refuses, explaining that the overall amount is too small for a deeper discount. All right, you got the best price you could. You instruct your financial planning module to conduct the necessary transactions. Meanwhile, you see that, based on best projections, your retirement shortfall has closed by four points, whereas the risk of achieving your planned nest egg has increased by four points. Somehow, it seems, all these financial instruments are adjusting too rapidly to one another. There is no free lunch, and you can't seem to close the gap without increasing risk. You know it has always been that way, but seeing it so crisply is a bit frustrating. Still, there is always room for surprising discoveries and new markets for your newly acquired stocks. That will surely break the risk-reward brick wall. Enough planning. You'd like an hour for a little lounging yourself before the day ends.

Much of this kind of financial planning is already happening with software packages like Intuit's Quicken, which can help you plan for retirement or for college tuition, transfer funds between accounts, pay checks from your computer, and do other activities that together are commonly called online banking.

Automated advice on what to buy and sell is not yet in place with home programs. But there are hordes of specialist programs run by investment advisors that can analyze and enact automatic buy, sell, hold, and change of status instructions based on movements in the price and volume of a stock being traded. Usually these numerical triggers are augmented by human judgment before the recommendation is made to actually buy or sell a security. Automating the latter will happen increasingly as such programs begin to consider for each security they examine the latest news—whether it's classified as good, bad, or neutral and whether it concerns mergers, profits, or bankruptcies. Today investment brokerage firms, like Merrill Lynch, maintain a large number of investment specialists on hand. Codifying their statements is straightforward and would most likely involve the hyper-organization middleware we discussed in the previous chapter.

Finally, the negotiating modules are eminently feasible and act not unlike the pipe managers that, as we saw, are skilled at negotiating transport fees. Except for the video bandwidth issues we have repeatedly discussed, this "future" scenario is feasible today.

A Virtual Compassion Corps

Over in Europe Grandma has a little more international communicating to do. A real night owl, she isn't bothered by the late hour; besides, it's the best time to reach her friends back home. The only part of these techie get-togethers she dislikes is the video imaging of the participants. She never had to get pretty for the telephone. Furthermore, her friends might detect her worsening posture and get worried. The post-polio is acting up again. But on this call she won't have to keep up appearances. She is contacting a post-polio group she recently joined.

Doctors aren't convinced that post-polio is a genuine medical condition, but there are enough people who had polio as kids and are now suffering that it doesn't matter what the doctors think. The post-polio victims around the world have found one another thanks to the Net, and they have organized themselves into a formidable group that collects and shares information on the affliction. The group even boasts some renegade doctors in its membership. Grandma shares her symptoms—difficulty breathing and walking—and learns that everyone else seems to have similar complaints. They also share suggestions for relieving the discomfort. More important, they can talk to like-minded souls who understand them and empathize. When will the medical establishment grow up and consider the qualities of empathy and understanding mandatory requirements for the practice of this noblest of professions?

Discussion groups for people with special needs have already formed on the Internet. Today's participants use typed text as their medium, but it won't be long before they use voice and images and eventually video. The real benefit of these groups is in how they link people with like needs—the community imperative that started it all back in the time-shared era. In few other ways is the Information Marketplace's power felt as intensely as in this ability to aggregate like-minded people across space and time who would not otherwise be able to reach one another.

This same power was also reflected in the exchange between Nicholas and Mbugua and could be evoked by national programs that pair children across racial, gender, wealth, class, and other distinctions. This is not imaginary folly: I have talked to Vice President Gore about urging U.S. college students to link up with "net pals" in poor U.S. districts and in developing countries, to help them enter the postgraduate world. This would involve discussing future plans, suggesting possibilities, and even arranging for more practical actions like setting up interviews with prospective employers.

This idea can be further extended to encompass a larger group of people and a broader spectrum of help.

Let me use this book to propose that the industrially wealthy countries of the world form the "Virtual Compassion Corps," an international alliance like the Red Cross or the Peace Corps that would operate through the Information Marketplace. At a minimum, this organization would act as a broker between people who wish to help and those who need help—a gigantic collection of "help offered" and "help needed" ads with mechanisms for matching suppliers and consumers of different kinds of human help. Such an organization could also help expose and reduce the fraud of intermediaries, who seem to capitalize on human pain and steal precious supplies and money destined for good causes. It could go further to establish much-needed larger projects and help them materialize.

This kind of help could be integrated into the world's welfare systems. And it wouldn't always have to go from rich country to poor country. Imagine a Sri Lankan doctor who earns very modest fees compared with those in the United States who could offer health care to Americans who cannot afford the high cost of care in their own country. They could plug themselves in to monitoring devices on medical kiosks at their neighborhood clinic, built by benefactors, and the Sri Lankan doctor could observe their vital signs, ask questions, and instruct a nurse to administer much-needed care, all for a minuscule fee, which might also be paid by a benefactor or a welfare organization. Imagine being able to offer this kind of service to the poor and homeless of America who now go without health care. In the other direction, retired U.S. doctors and medics who want to help people at home or overseas could join teachers, farmers, and numerous other professionals who would offer their services, free of charge, to the needy. They could do so for an hour here and an hour there from their homes, achieving a great deal of good around the world and much personal satisfaction.

I can already hear the protests about mismatched medical standards, abuses of the welfare system, and more. Where there is a will, such problems can and should be overcome. For example, a local organization could be set up to assume liability for the work carried out by distant doctors. But let's not quibble about such details. Instead, let's consider a world where people on a free-will basis use the Information Marketplace as the great matchmaker between those with a natural willingness to help and those with a natural need to be helped.

That could make a difference!

Your Day

Following a fictitious family around on one future day touches on only a very few ways in which the Information Marketplace will change our daily lives. To get an idea of the scope of changes to everybody's lives, imagine the many different kinds of families and people there are, all with different interests and habits. Some may prefer to follow sports, participate in PTA meetings without leaving home, have a monthly class reunion, repair their cars and washers and other appliances with remotely proffered advice, hang with fellow teenagers, flirt, look for adventures, find and certify a babysitter for an upcoming party, find work, engage in a part-time job like selling real estate, check on the health of the house while traveling, plan a trek, weave colorful blankets with a $500 computer-driven loom attachment, play poker or bridge with friends who have moved away, and on and on.

Pick a typical day of yours. With all you've have learned about the Information Marketplace, try to project yourself doing similar activities a decade or two from now. Think of the new artifacts we have introduced and the interfaces that will be around you, the services and shared tools that will be available to you. You'll then be writing this chapter for yourself.

6 Pleasure

From Your Easy Chair

Eating, shopping, driving, and visiting rank among the most common activities in daily life (we'll get to our work lives in chapter 9). When we're not pursuing our routines or obligations, we seek pleasurable activities. Indeed, for many people, the very purpose of work is to provide the means for leisure and pleasure. Whether you work to live or live to work, you no doubt care about your pleasures and the ways they will change in the Information Marketplace.

As with daily life, innumerable activities will be affected. We'll focus here on a few of the more common pleasures—watching movies, having sex, taking pictures, engaging in hobbies, creating and experiencing art, playing games—in order to see the wide range of changes that might be wrought by the Information Marketplace.

Entertainment will be the first aspect of our lives to be dramatically affected, because people crave it, because there are more TVs than computers in the world, and because companies see enormous profits in delivering immediate access to every movie ever made and every concert and song ever recorded. This will be the first major commercial use of the U.S. Information Marketplace. Experimental applications of many of the ideas we discuss in this section are already under way in some two dozen communities of consumers.

Large media companies are making big bets along these lines, given the promise of high returns. There are nearly 150 million TV sets in the United States, owned by people who already enjoy watching lots of movies. In Europe, where cable TV is unevenly distributed, and where there are some pockets of protest concerning cultural invasions from America, the evolution of the Information

Marketplace may start differently—from education, cultural events, news services, and other erudite sources—at least, if European policymakers are right in their assessment. That is not likely. To this Greek-American author who has commuted between the two continents for over twenty years, the press and the European political theaters make much more of the supposed differences between Americans and Europeans than is actually the case. People are people, and when they are able to make their own choices, they are more likely to act alike than differently! One need only look at the consumption of blue jeans, fast food, and pop music in Europe—or of Swiss chocolate, German beer, and British rock in the United States—to arrive at the same conclusion. Let this be a bet that throughout the world, information infrastructures will exhibit many similarities, just as worldwide television does today.

Movies-on-demand will be here within a few years, as soon as the pipes can support video. On your TV screen you will navigate through a catalog of thousands of movies. Or you will browse through an even larger list of all the songs ever recorded—by Madonna, Pavarotti, Edith Piaf, Smashing Pumpkins, the Grateful Dead, Nat King Cole, and every other artist and group in the world, including aspiring amateurs whose songs, mercifully or regretfully, will be free. All the symphony orchestra performances will be there too, next to all the musicals, the *Nova* and *National Geographic* films, the how-to videos, the operas and theatrical performances, and on and on.

You will explore from your easy chair, first with your mouse and later with spoken dialogue, while watching your screen and listening to verbal responses. Within a decade, commands like these will be commonplace in the U.S.: "Let me hear this morning's 'BBC World Service.' How about a little Mozart now? Jazz please—something I've really enjoyed before. How about that movie with Marilyn Monroe where her skirt blows all over the place? Let's go after the Bach cello suites by every cellist who has ever performed them . . ." You might prefer the more structured method of browsing over a menu of options—dramas, comedies, action films, classics, family films, foreign films—narrowing the search with the names of actors and actresses you like, or the favorite choices of critics you resonate with, or the most popular requests from other viewers like you. Or, as we saw with the wakeup music, you might supply your movie rental service with reactions to the movies you watch so that it can suggest others you will probably like.

Regardless of how you make your selection, you'll sit back, relax, and enjoy the experience that jumps to your senses in brilliant color from your living room wall screen, or your goggles, with realistic video and concert hall sounds. The cost of such a service by the early 2000s will be competitive at a few dollars for a full-length movie and a few cents for an hour of music, charged automatically to your account. The prices will come down from the mid-1990s experiments as a result of intense competition among the information content providers and a much larger and more international audience than the video rental stores could have ever imagined.

Once the movie or musical is over, you contemplate making a copy of it, because you would like to show it to your best friend who is out of electronic reach exploring a cave. But the cost of copying is so close to the rental fee that you simply e-mail her a note about where to find the movie when she has time to watch it.

Besides e-mailing movies and songs, automatization middleware will be used to execute your standing instructions about how you want to be entertained. Pipe managers will strive to find the needed bandwidth at the lowest price. Speech-understanding modules will let you hold a dialog about what you want to see and hear. Security schemes will ensure that children don't have access to material you don't wish them to be exposed to. And groupwork modules will be invoked to let you interact with other people and your entertainment content.

That last point warrants a comment. We've heard a lot in the last few years about how the new videos in the Information Marketplace will be fully interactive, giving you some control over the plot. "The murder has just been committed and the killer is running down Elm Street. The police are close behind on foot, trying, like you, to find out why he did it. Do you want him to have killed for passion, money, or fame?" Make a choice, and the movie continues in a certain direction. This choose-your-direction approach saves time when using a how-to video on repairing your garden tractor; you pick the question and get an answer instead of having to wade through the whole movie until you find the place that interests you. It will also play well in virtual sex, as we will soon see. But you may not feel compelled to interrupt and redirect the movies you watch. As we'll note in our discussion of art, there are good reasons for the artist to retain control and for you not to want to make decisions while you are relaxing and being entertained.

Forbidden Pleasures

In the early 1980s as PCs were beginning to appear, I was interviewed about the future of computers on the *Christian Science Monitor* TV program. With thirty seconds to go, the interviewer asked me pointedly what might be the most outlandish uses of computers I could imagine. I replied, "Tele-food and tele-sex." When the ruckus quieted down, I was asked to explain. Fortunately, the show was over, because I had no idea what I would have said, beyond the obvious and bizarre definition of having food or sex at a distance and the huge technical difficulties of achieving either of them.

Tele-food was impossible then and will remain impossible until, if ever, we figure out how to fashion biochemical structures with computer commands. However, tele-sex, as I called it then, is a lot closer. Fifteen years after that interview we could fill several books the size of this one with stories on cyber-sex, as it is often called now. Instead, let's try to analyze the realities of this age-old activity in the brand-new setting of the Information Age.

Just as the Information Marketplace will allow you to dial up a movie or a song, it will allow you to dial up services offering sexual experiences. Because there are only machines and humans in the Information Marketplace, there can only be two categories of sexual activity—human to machine, and human to human, each mediated by information infrastructures. The machine-to-machine category is of no interest: enlightened people would hardly worry about what consenting machines might do with each other!

At its simplest level, human-to-machine sex is the electronic equivalent of thumbing through a sex magazine. In place of the magazine there is a computer, a "server" in techie language, that is stuffed daily with new titillating images and videos. In the mid-1990s such bulletin boards were beginning to appear, confined to still images and organized along subcategories of sexual activity. Clients would pay on the order of $100 a year for the service, which they could access at any time. Robert and Carleen Thomas, a husband and wife team who ran such a bulletin board, got in trouble with the law in 1994. Images that were legal in California, where they and the server resided, were accessed by a postal inspector in Memphis, Tennessee, where the commerce in such images was illegal. The Thomases were arrested, extradited to Tennessee, tried, and convicted. He was sentenced to two and a half years in prison;

she got three years. We'll delve into these kinds of cross-border dilemmas toward the end of the book. For now, let's just observe with some certainty that, given human nature and history, this kind of sex hyper-magazine service will grow and migrate to video when the information infrastructure's bandwidth can support video exchanges that are better than today's jerky, grainy flicks.

If you look for the next level of technical, if not sexual, sophistication on today's information systems, you will most likely encounter an interactive video that you control. It may reside on a remote system that you access through the information infrastructures or on a CD-ROM you insert into your computer. A scantily clad human model fills your screen and says, "I am yours. What do you want me to do?" You respond, "Take everything off." The model obliges, moving seductively in the process, and then asks for your next command. Early versions are restricted to a fixed menu of commands selected with your mouse. It doesn't take much guesswork to forecast that future versions of this genre will surely utilize spoken dialog, increase the range of possible commands and reactions, and offer additional features like repeating your first name for greater personalization.

These activities are controversial. For example, even though the above description was crafted to be gender-neutral, many readers will read it as men exploiting women. Others will see it as an immature yet safe form of sex. And still others will see it as an opportunity to invent more creative variations. Whether you approve of such things or not, when you encounter them in the Information Marketplace you will no doubt feel toward them as you feel today toward sex magazines and videos.

This is as it should be, because these breathless "innovations" are nothing more than straightforward extensions of current media. The use of interaction does not change the nature of the fundamental sexual exchange, which is between a human and an artifact. And as long as there is an artifact rather than another human in the picture, the effectiveness of the medium and the pleasure of the participant will vary—in roughly the same sense that sitting under a tanning lamp watching a beach movie might compare to actually going to the beach.

My calling such sexual exchanges *straightforward* extensions of current media will be seen as an insult by some technologists and entrepreneurs. They will crank up their inventiveness, providing full-immersion video, goggles, haptic gloves and bodysuits, and many more unique artifacts that might be coyly described as exotic human-machine interfaces of the tactile variety. I can hear them now: "A sexual experience can be arranged with a physically present humanlike robot or with a virtual partner that you can feel in every respect through a bodysuit. What you see, feel, and hear can be made so believable, so close to reality, that it will silence your stupid criticism. Open your eyes, man, and take in the new world. You might even like it!"

We'll talk more about such possibilities shortly, but don't hold your breath for this vision to come! Leaving aside the immense technical problems and the consequences for your body of a computer crash or a malfunctioning motor in your bodysuit, the prospects for this kind of pleasure are dim. Like the infamous rubber doll, even a sophisticated physical or virtual artifact that strives to simulate reality remains always an artifact in its user's primary sexual organ—the brain. At a deeper level, even if a twenty-first-century robo-doll were to become realistic—moving, breathing, and talking flawlessly (a huge technical "if" that will not be solved anytime soon)—it will still exist behind a profound barrier that separates the real world from the informational world, a real human presence from a virtual human presence. We will discuss this at the close of the book. Certainly, novel sexual artifacts will appear, but ultimately they will be relegated to curiosities for the few, like today's rubber dolls.

On to the second category: human-to-human sex mediated by the information infrastructure. One of the obvious and early services involves a live human model at the service provider's studio, standing in front of a video camera, interacting with you visually and verbally, wherever you may be, obediently executing your requests for a fee, which in the mid-1990s was around fifty dollars for a fifteen-minute session.

This setup is similar to the CD-ROM striptease, except for a very big difference—a real human is at the other end of the connection. In that sense, this activity is similar to phone sex, which was made possible by special telephone numbers and novel billing mechanisms. The video experience we describe here does not require a computer at all. In fact, if we had adopted videophones after the invention of TV, this "new" application could have arrived almost half a century ago! So technically, the idea is old hat. Psychologically, it will be perceived as better than CD-ROM sex because it involves a real person. Yet, like phone sex, which may provide some basic excitement, this form of pleasure lacks the sincerity and genuine involvement of the paid human partner. Nevertheless, because it uses video in addition to audio, it will attract customers away from phone sex and entice a new set of customers. So, expect video sex to become quite popular.

If you're keeping track of our progression, you will have already guessed that the next stage of human-to-human sexual activity involves full-immersion interactive sex at a distance with another human—a willing partner. Before I describe the setup, let me assume that you have swallowed hard and decided to try it. It can be done in several ways.

In the first approach, there is a physical robo-doll at your partner's site that represents you. You wear a virtual reality bodysuit that detects all your movements and causes your robo-doll clone to repeat them at your partner's site. The clone is equipped to transmit back all its sensory experiences—all the sounds it hears to your ears, all the sights it sees with its camera-eyes to your goggles, and all the tactile sensations that it feels to your suit's effectors.

Alternatively, both partners can wear full-immersion sex suits and throw away the robo-dolls. Your actions are felt by your distant partner through his or her suit, and you feel your partner's actions through your own suit. Because the linkage between the two suits is pure information—numbers—the mediating computers could process this information to create some interesting and bizarre effects. Participants may chose different views of their partner, in a strange variant of the mirrored ceiling. Or they may elect to appear through their avatar with a different face, longer hair, broader shoulders, perhaps as humanoids from a strange planet sporting extra limbs . . . and worse.

Today, we accept cosmetics, tailored suits, plastic surgery, and many other products and services that make us look more attractive. Indeed, we have been purposely altering our appearance for more than 6,000 years. So what will stop us from using bodysuits and virtual reality, which are guaranteed to improve our looks far more dramatically?

More bizarre encounters are possible. In a mixture of Roman gladiators and French salons, two robo-dolls (physical or virtual), representing their owners, may lie intertwined on the floor executing the commands of their masters while the latter sip their martinis and observe. Or, a famous performer may be cloned to a multitude, giving a few thousand people the opportunity to experience via their bodysuits the performer's sexuality—simultaneously.

New questions and variations of old ones will arise, too. Will it be considered untoward behavior to have sex on the first date, or even before any date, if it is done through the antiseptic separation of virtual reality? Will the millions of youngsters who spend their nights dancing with explicit sexual gyrations in clubs, or watching such dances on the music-video TV channels, refuse to engage in virtual sex because it is immoral? If they succumb, will their parents, teachers, and clergy be upset? Some of these older role models subscribe to the dictum "You may watch but don't touch." Is a virtual sexual encounter closer to watching or touching? We have trouble answering such questions from our twentieth-century perspective because of the disorienting blending of the virtual and physical worlds. But sooner or later we'll have to tackle them.

When we do, chances are that human intent and how we want to feel about ourselves will take precedence over these fascinating but more mechanistic quandaries, with the result that we will opt for the same mix of moral positions, noble human values, and their less lofty opposites that we have maintained for thousands of years.

How people interact with a person's "avatar," or virtual presence, versus their actual presence is certainly a new relationship to be pondered. It has nothing to do with sexual encounters per se, and applies to all virtual encounters between people in the Information Marketplace. Throughout history, the public has always shown the servants of powerful people some of the respect accorded to their masters. In the same sense, people will probably extend to your avatar some of the behaviors they would extend to you. This became vividly clear to some of us in 1995 when Vice President Al Gore visited our lab. We gave him a chance to drive Rover. Developed by Dr. David Tennenhouse and his students, Rover is a toy-sized remotely controlled car equipped with a small video camera so that the controller sees on a video screen what Rover "sees" in front of it as it zips around. The car can be driven from the Web, which means that anyone—from nearby or a thousand miles away— can drive Rover up and down the corridors of our lab, enter an office, look up with its camera at who's there, and hold a conversation using its speakers and microphone. Rover is most definitely an extension of the person driving it.

When the vice president took the controls, he drove Rover out the corridor and approached a barrier set up by the Secret Service to separate secure from insecure areas. No one was supposed to go beyond these barriers. As the agents saw the car controlled by the vice president approach the barrier, they were clearly disturbed. You could hear their minds clicking: "This thing is an extension of the vice president, and it's approaching a danger zone. Is the VP in danger? Never mind. We can't let it through." One of the agents abruptly stepped in front of the little car and made wild gestures, trying to signal it to stop and turn back. Upon seeing all this commotion on his screen, Mr. Gore, several rooms down the hall, let out a charging yelp and, laughing uncontrollably, swerved expertly around the agents and past the forbidden barrier to freedom!

Back to virtual sexual encounters: even if we are prepared to accept them, we may be in for a long wait. Computer-driven dolls and suits for human-to-human interaction are not yet here. Cheap and crude varieties will no doubt appear and find their small niche with the adventurous set. But the deluxe and totally safe models could take decades to develop and are likely to face substantial technical and expense barriers.

As for the human aspect of these activities, the best we can say is that when willing and loving partners are involved, the interaction has the emotional possibilities and human dimensions of phone calls and letters between loved ones. Taking our cue from these older forms, we can predict that virtual sex will be seen as inferior to face-to-face contact. And yet, this new kind of sexual engagement could still be useful in situations where husband and wife are physically separated for long periods of time as a result of work requirements, prison sentences, or military duty. It could be even more useful and humane for people with handicaps that leave them unable to engage in real-life sex. And through augmented or virtual reality, we might even achieve new forms of sex instruction. So we should not be too quick to thumb our noses at it.

Just as we can foresee new variations on existing sexual stimuli, we can foresee new variations of existing problems related to human-to-human sex via the Information Marketplace. One example stems from current interactions based on anonymity and impersonation on the Internet: a young male pretends to be a woman, an old man parades as a teenager, a physicist takes on a laborer's persona, and so on. These sorts of acts, while fashionable today, will wane as anonymity becomes less feasible and less acceptable. People are not born with an inherent right to anonymity, especially if they are intent on disturbing somebody else's peace. People will not insist on remaining anonymous either, any more than they do in today's society. The Information Marketplace will simply reflect the same human social habits. If you don't want interactions with anonymous or pseudonymous people, you won't have to deal with them. As we saw in our security discussion in chapter 4, the infrastructure can guarantee that such characters are kept away from you, for example, by your requiring everyone who wants to communicate with you to use their private key, thereby forcing them to disclose their identity (when you use their public key at your end). If you like to visit bars and enjoy the adventure of anonymously meeting new people, the technology can easily be adjusted, or dropped altogether, to give you the option of doing so.

Our discussion so far presupposes consent if not love among participating adults. This will not always be the case. Though you can always turn your machine off to avoid virtual sexual abuse, the potential for criminal activities is there, especially where it involves the luring of minors. We will address computer crimes later in our discussion of legal and other means for controlling access in the new world of information. With the bad encounters, we will also hear increasingly about Information Marketplace friendships that turn to wonderful marriages. The same human nature, good and bad, that finds its way into every city of every country of the world will also find its way into the Information Marketplace. Although the anonymity and proximity possible in the Information Marketplace seem ideal for undesirable activities, the medium will encourage no more or less a basic shift toward the bad or the good than mail and telephones did. So a certain degree of calm and introspection are warranted, in lieu of overreaction or indifference. And as in real life, for every measure a countermeasure will always be possible, with both measures and countermeasures becoming increasingly sophisticated. Ultimately, looking for goodness or guarding against evil is no easier in virtual life than in real life.

Creations and Flashbacks

The Information Marketplace gives us new mechanisms and interactive control to pursue ancient human drives like sex. It will also give us new kinds of creative controls for less complicated and less ancient forms of pleasure that help us capture and create sensual experiences.

Digital cameras are already available. They capture images by recording the lists of numbers representing the color pixels, as we discuss in the appendix. They can then feed this data directly into your computer, and you can call these pictures up on your screen. The image detail is not yet as refined as printed photos from conventional film cameras, but it will get there within five years. Printed photographs can also be scanned into a computer with a scanner. Drawings can be scanned or created on the computer, too.

The fun really starts once these images are in the machine and you begin to transform them. Even today, programs like Adobe's Photoshop can be used to change the eyes of your portrait from brown to a lurid green, to remove a blemish on your left cheek, or to cut a friend's picture from one photo and paste it next to his enemy, crafting the two of them so they hold hands. You can also edit and blend human- and computer-generated pictures, limited only by your imagination. More sophisticated programs, like the ones Mary used to design her theater, are available for architectural uses. They convert drawings into colorful perspectives and even three-dimensional models that you can visit via your screen. These fancier programs are still crude but will improve as computer power increases in the next decade. All this software and hardware will trickle down to the consumer at affordable prices and will be available for all sorts of fun and games. After a couple of decades, digital photography will have replaced chemical film photography as surely as that older craft replaced portraiture.

The same things can be done in people's homes with digital video, at greater expense, as are already done in professional video studios and TV stations. Meanwhile, consumers will embark on a new binge of digital audio and music making of their own, because that machinery is potentially cheaper than its video counterpart and all of it is already in the hands of the professionals who produce CDs. They use computer editors to stitch a perfect CD like a quilt out of many imperfect takes—manipulating the duration and pitch-bending of each and every musical note or even fraction of a note!

Images, CDs, and videos are familiar media that will benefit directly from the Information Marketplace. Everyone will be able to create them, alter them, and move them around. The simple act of e-mailing these new creations will give new meaning to the sharing of mementos among friends and of commercial information goods among buyers and sellers. At the same time, the ease with which this information can be changed will lower the believability of photographs, recordings of conversations, and even videos. Unless they are digitally signed by a trusted party, tomorrow's photos, videos, and sound recordings will be regarded with the same respect for authenticity we would accord today to unsigned typed text.

As we saw in our discussion of new e-mail tools, the Information Marketplace will also provide a whole new way of recording and sharing human experiences, like the swearing-in of a president, or a sky dive from a plane, or the arrival of a tornado. You put on your goggles and bodysuit and "feel" the entire event as if you were there amid all the hustle and bustle. Even though these kinds of recordings are a decade or two into the future and may be grainy and imperfect, they will nevertheless come our way.

We have no idea how far these more fanciful forms of pleasure will take us. But we can be sure that the entire audiovisual home experience, whatever else it does or doesn't do in the Information Marketplace, will be one big blast of fun.

Art

The manipulation of photography, video, and music brings us to the world of art, the creation and appreciation of which is one of humanity's oldest forms of pleasure. It too will change.

Though most artists profess to be at odds with technology, they have consistently used the latest tools to create new artistic processes. Throughout history, art has evolved in step with technology. Noise-making sticks led to drums and harpsichords and electronic synthesizers. Cave frescoes turned into paintings and then photographs. Clay pots became marble statues and neon sculptures. Ancient theaters developed into music halls, cinemas, and video rooms. Because we conceive and perceive the pleasures of most art through visual and auditory processes—strong suits of information technology—the Information Marketplace will bring several radically new dynamics to the creation and appreciation of art.

In Greek mythology there is a story about Prometheus. He stole fire from the gods on Olympus and gave it to mankind. As punishment, Zeus chained him to Mount Caucasus. Each day a vulture would fly down to tear at Prometheus's liver. In a mythology book you would read the author's hair-raising description of Prometheus's anguish as the vulture torments him. In a movie you would see the vulture approaching from the sky and hear its frightening screech. In the Information Marketplace, you might put on your virtual reality helmet and haptic bodysuit and experience the event fully from Prometheus's perspective:

Your limbs are chained to the mountainside. You struggle to pull free, but you cannot. The rock wall feels cold against your back. The evil black vulture appears in the blue sky above, circling. It turns and descends swiftly toward you. You kneel in fear, then crouch down on the hard ground, straining against the chains as the vulture's huge wings spread just above your head, blocking out the sky. You instinctively cross your arms before your face. The vulture's thrashing wings beat against your head, flapping loudly, as its sharp beak playfully jabs your side; you wrench your body away in pain. Suddenly, the giant bird rears back its big head, its wild eyes widening in preparation for the deep, piercing strike. You let out a primal scream: "Noooo!" The vulture screams back, "Say the magic word!" and remains motionless above you.

After your heart stops pounding, you say, "Great art!" The vulture smiles and morphs into your daughter's face.

You have known all along that the vulture was really your daughter, twenty miles away in her sixth grade classroom, wearing a special bodysuit as she impersonated the evil creature as part of her class project on mythology. The Information Marketplace translated her movements and commands into what you saw, heard, and felt at your office. Yet even knowing that, you have not quite stopped shaking as you congratulate her on her creative performance.

Hard to imagine? Not really, at least in the long run. The head-tracking helmets are already in use at premier virtual reality labs such as that of Henry Fuchs at the University of North Carolina. And, as we have seen, several researchers are working on goggles and haptic interfaces. Thus, one new dynamic the Information Marketplace will bring to art is *the involvement of several senses and muscles* through visual and auditory immersion, haptic interactions, temperature changes, and controllable smells. As these systems improve, artists will combine these different sensations to create novel and fascinating experiences.

We may encounter the resulting new art forms where we least expect them. One scenario came to my mind after I visited the home of my colleague Nicholas Negroponte, now head of MIT's Media Laboratory, in the early 1980s. I found no place to sit inside his living room; every couch and recliner was occupied by life-sized, human forms made of white cloth by a local artist, Myra Cantor. On my way home I envisioned three of these forms, two men and a woman, sitting in chairs in the reception area of our laboratory—all wearing business suits. They would have motors in their joints, audio speakers behind their mouths.

Now imagine it's an average Thursday morning in May. A representative from a high-tech firm along Boston's Route 128 walks in for a meeting with me. A bit early, he sits in an open seat and cracks a smile as he notices the motionless human forms—another hack of the crazy lab people, he muses. But imagine the expression on his face when the female statue next to him suddenly clears her throat, the male across from him lets out a gentle sigh, the other male crosses one leg over another . . . then all three of them change position to match the way the visitor is seated. Suddenly unnerved, the man shifts in his chair. The figures do the same. He peers at them nervously; they peer back. The figures mimic the man's every movement and sound.

All this, of course, would be done the easy way by several of my grad students secretly observing the visitor, each controlling a mannequin to create the maximum level of discomfort for the visitor.

I found an artist crazy enough to pursue this project but was stopped from pulling it off by MIT's accountants, who cited constraints on how we could spend our research funds. So I put the idea aside. Perhaps a technically minded artist will be inspired to implement the gag.

If so, will it qualify as art—perhaps a new category of performance art? For that matter, would the Prometheus scene be considered better art than the book or the movie? The questions are silly. Just as books and movies are each better suited to convey different artistic messages, the new art forms made possible by the Information Marketplace will find their niches.

Another new dynamic the Information Marketplace will bring to art during the next decade is *interactivity*. Interactive plays, in which the audience determines the plot, have been around for a while. But the Information Marketplace will give rise to many alternatives, and some gems are sure to arise.

Imagine yourself seated at your piano at home. You have electronically paid for a special treat that is about to begin. You are wearing two of the haptic gloves described earlier in the book— gloves laced with actuators that allow a computer to move your fingers and sensors that transmit every fine motion of your fingers back to the computer. There is a video screen at the end of the piano, high-quality speakers in the walls. At his California home Alexander Borkin is wearing a passive pair of gloves that sense his hand and finger movements. Borkin, our imaginary popular interactive pianist, will be giving an experimental performance.

Borkin places his hands on the keys of his Steinway. Your gloves, directed by his, position your fingers in the same way on your keyboard. Borkin begins to play a melodic polonaise. Your gloves reproduce his exact hand motions, and your fingers begin playing the keys of your piano. You hear the building tones. Borkin strikes the keys harder; your hands do too. He eases off; so do you. You never knew "your" hands could make such music!

The polonaise gives way to a modern piece. A few more gadgets allow Borkin to deliver uncommon whistling, humming, and exploding sounds to your room simply by moving his head or torso or eyes in addition to his hands. Correspondingly uncommon video images flash and swirl on your screen. You feel the swaying, finding yourself engulfed in the audiovisual experience. Quite a few experiments have already been tried with dancers instrumented in such ways. What might Borkin create, and what might you experience on the other end?

So far the artist's creations have been controlling you. Now it's your turn to contribute. You take over control of the video images and accompanying sound effects. There are no rules. You reach up in the air with your gloves and move your arms; you tap a double beat with your forefinger; then you push down harder with your gloves, trying to complement the piano playing. Patterns, colors, and sounds change to suit your actions. Together you create a fascinating musical variation while simultaneously creating new video splashes.

Meanwhile, in the next room, your spouse is standing before an easel, immersed in a similarly interactive creation with a painter at her canvas in Seattle.

Margie, your performing artist friend who is witnessing all this, recoils in indignation. "That's not art. If you are just repeating his mechanistic movements without experiencing the feelings that lead him to act as he does, then you are no more than another one of these programmable machines." Perhaps so. Then again perhaps not. History teaches us to let a few centuries go by before conferring judgment on new art forms.

The world's research laboratories are likely to make these scenarios a possibility, if not an immediate commercial reality. Recall our earlier mention of the haptic interface called the Phantom. It senses the position, configuration, and forces of your finger, while resisting your finger movements as directed by the computer. The technology for the full-glove version of this finger is feasible but ten to fifteen years in the future. The other aspects of the Borkin scenario can be handled by the shared tools we discussed in chapter 4.

A third new dynamic in tomorrow's artistry adds a variation of groupwork, which we'll call *group play*. Now with your movements you attempt to persuade Borkin to play louder, use a wider range, create even more daring video imagery with more pronounced body movements. Other participants like you fitted with gloves in other locations and time zones do the same, trying to cajole the artist along preferred directions. Borkin, now wearing haptic gloves and a haptic jacket, feels a scaled-down and cumulative version of all these nudges by the audience. Occasionally, the participants resonate with one another, cueing the artist uniformly toward their collective whim. When Borkin shares and amplifies these feelings, a new and strange art form emerges based on the bonding feedback generated by the intertwined ensemble of artist and participants. Fantastic!

As promising as interactivity and group play sound, they may not turn out to be popular. As we've discussed, after some initial excitement and experimentation, people may realize that they don't want to make decisions when they are being entertained or when they're experiencing art. I can almost see the e-mail flame of the year 2002, following a much touted interactive movie: "Don't patronize me with this false sense of creativity. Either give me all the tools I need to make my own first-class movie from scratch, or stick to delivering POMs (plain ole movies)."

Artists may not be compelled, either. As art becomes interactive they may find themselves giving up control of the creative process, compromising their ability to make an individual statement. Other artists, however, may see a new opportunity for making precisely such an individual statement in the way they orchestrate the interactions. Who knows; group creation might become all the rage.

The final dynamic the Information Marketplace will bring to the creative world is the *democratization* of art. It may not be the most exciting development, but it will be the most important. Suddenly, all the world's art will be available to all the world's people. Already, the world's museums are putting their most popular works on CD-ROMs and on the World Wide Web. You can experience symphonies and plays over the Internet. More online art will come from media companies, galleries, universities, and other institutions that want to show off their wares to the largest possible audience for professional, commercial, or pride-related reasons and in hopes of finding new patrons. So will the world's symphony orchestras, operas, music ensembles, ballet companies, theater groups, and the many individual writers, poets, painters, sculptors, composers, singers, and performers.

Neophytes will also want to expose their creations to the global audience. The costs for mounting an exhibit on the Web will be from 10 to 1,000 times cheaper than today's cost of renting a small gallery or exhibition hall.

For a long time, my dream about my native Greece has been to put its cultural treasures in the Information Marketplace, giving everyone the capability to access them with a fancy historical fly-through. It will be captivating to be able to view this ancient civilization by "steering" our virtual "histori-copter" through space, "diving" or "rising" through the ages, and using colors to identify the different kinds of endeavor, such as political events, natural events, writings, sculpture, pottery, music, folk customs, and dress. Each of us could visit this old world, the root of Western civilization, in the way that we find most to our liking. Seasoned historians are able to do this kind of "flying" inside their heads, but for most of us it is completely out of reach unless we have such an easy and inviting historical vehicle under our command. Usable versions of such explorers should be around by the turn of the century.

These developments will bring the ability to view, experience, and create art to many more people. If you believe that being exposed to more art and being able to create more art are good things for the world, then you will find this proliferation beneficial. If, on the other hand, you have been browsing the World Wide Web lately, experiencing the writings, illustrations, and other "art forms" that have evolved on it, you might be tempted to regard many of these creations less kindly. You will then actively seek the publishers and intermediaries of the Information Age who can help you find the diamonds in a mounting pile of info-junk. Less democratic, perhaps, but more enjoyable. Still, whatever your take on what's available right now, you're likely to welcome the Information Marketplace's ability to bring you art from around the world at far less cost than is required today. You won't have to fly to Rome to take in the Colosseum, or slog through the mud to experience Woodstock II. Displays at the Louvre in Paris, performances at the Bolshoi Ballet in Moscow will come to you. "But," you object, "nothing can replace the real-life experience of actually being there." Take a moment, and look around your living room:

Lunch is almost finished when the videophone rings. A large section of the wallpaper dissolves into the live image of your newly made friend who shares your love for music—the woman whose dog you found. She's adorned in a stunning scarf, and she's inviting you to attend a virtual-reality-enhanced concert and instrument demonstration by the Vienna Philharmonic Orchestra at the Grosses Festspielhaus in Salzburg, Austria—in fifteen minutes. You have no plans for the evening, and there is plenty of time to put on your virtual reality glasses and bodysuit and ask your living room computer to link you with your friend and to then move you with her through the concert hall. She has made provisional arrangements already. So you find yourself talking to an automated ticket agent in Salzburg, where both of your tickets are waiting. One electronic cash transaction later, and four minutes since her call, you "walk" into the concert hall with your friend at your side.

The walking that you do, of course, is on the floor of your house. Your home computers know the layout of your rooms and furniture and have reconfigured what you see as aisles in the concert hall and virtual ushers and other concertgoers, so you move though your house without stumbling into anything. A familiar smell penetrates this bliss, gently reminding you, as you bow to an acquaintance, that you must be walking through your bathroom! You make it to your theater seat, mercifully back on your living room couch, next to your friend's virtual presence. Together, you plunge yourselves into the delicious sounds and sights of the strings being blended by expert performers to ethereal voices in full splendor. During the performance the diva slips and falls down. But she is not one to shy away from innovation: preserving the pitch and color of the last note, she sings, "O how painful to collapse," and continues from a reclining position to recite the next aria from *Don Giovanni*. The audience applauds with appreciation, and you rejoice in your good luck at having witnessed this rare event.

During the intermission the performers come out on stage so you and other members of the virtual and real audience can ask questions and try out the many instruments being demonstrated. Your friend quickly says, "Try the viola d'amore." You reach for the richly colored instrument that will be used the following day in J. S. Bach's *St. John Passion* and pluck the lowest string, hearing clearly the overtones produced by the sympathetic string tuned one octave above the one you are plucking. What a crazy fourteen-string viola, where half the strings are there just to resonate. Your trial is not impeded by the few hundred people around the world doing the same thing with this and other instruments, for they are now hidden away by your computer in the interest of personal service. You can turn them on if you want, for a fuller effect, but it's no fun watching a hundred hands plucking away at the same instrument or hearing the din that results.

The chain of computers between you and Salzburg have done their job, letting the Festspielhaus's host machine know exactly where your finger was in your living room, then quickly translating this information to the corresponding position in the concert hall, which in turn was identified as plucking a certain string of a certain instrument. The Austrian host computer then issued back the resulting sound. You are oblivious to all these computer machinations as you marvel at the deep tone of this splendid old instrument. Soon, the second act is under way.

After the concert is concluded, you realize that the long day has taken its toll. You need sleep. Real sleep—virtual won't do! You kiss good night, at a distance, feeling nothing physical but lots of anticipation.

Though a couple of decades or more into the future, this kind of "being there" will bring a new splendor to our lives. It will lead to new art forms and new ways of enjoying the world's cultural heritage. As exciting as this prospect is, it is not essential for the Information Marketplace to have a major impact on art. When all is said and done, we may be surprised to find out that people will benefit more from the democratization of art than from the fancy new interfaces, interaction, and group play that the Information Marketplace will bring.

With or without new-fangled artifacts, the Information Marketplace, by enabling people worldwide to partake in the same experiences, will begin to create a universal cultural veneer, a common experience of sharing in humanity's art that transcends the differences between our geographic and ethnic bounds. I will have much more to say about this in the final chapter of the book.

Virtual Neighborhoods

The Information Marketplace will enable groups of people scattered around the globe to create and appreciate art and artistic events. What else might happen when groups set pleasure as their goal? One of the most immediate effects will be the formation of virtual neighborhoods.

A virtual neighborhood can be envisioned as a conventional physical neighborhood that links together a few hundred people in a small town, or a large city. In this case, though, the streets leading to and from these people's homes have a magical property: they can carry you instantly to a neighbor living in the next house or the next continent. More important, the similarities among the neighbors need not be limited to social and economic status, as is the case in physical neighborhoods. In the virtual neighborhood, they can expand to thousands of different dimensions.

Rudimentary versions of virtual neighborhoods appeared three decades ago on time-shared computers and persisted through the transitions to the Arpanet and Internet. Known as *bulletin boards* and now as *newsgroups,* these common interest groups bring together people who type their comments on a shared file for public consumption and response. Because of the accumulated knowledge, any questions get answered rapidly and in detail. Also, because of the large number of participants, the groups can sometimes drown in an electronic mountain of trash and flames. For some topics, like literary, political, or issue-centered discussion, today's newsgroups fare well, because the limitations of typing and delayed responses are helpful to the discussion process. But compared to most of tomorrow's virtual neighborhoods, with their sensory and effector interfaces and sophisticated groupwork modules, the current interactions are roughly like talking about a shared interest versus actively pursuing it.

Perhaps you collect stamps, coins, quilts, antiques, photographs, toys, or dolls, and you enjoy looking at them and trading them. Maybe you are interested in golf, skating, football, the decathlon, baseball, or soccer, and you want to share videos of (or later your presence in) such events. You may be into beer or wine making, woodwork, needlework, or any other of the world's crafts, and you may want to share with your virtual neighbors programs that drive your hand tools. Or you may want to play bridge, chess, or Maze Wars. Perhaps what you really like is poetry or comedy, and you want to share readings and performances.

Whatever your interest, you will be able to find people in the Information Marketplace who share it. Until now only big cities could offer this advantage, and even then in limited form. People with ordinary hobbies, let alone exotic ones, have had to work hard to find others who share the same interests and are within physical reach. You want to get together with them to show your goods and see theirs, to discuss relevant issues, swap ideas, and explore. The Information Marketplace removes the barriers. All you need to do is find a cluster of these kindred souls. Once you have found them, you have moved into a virtual neighborhood—with its own language, customs, and mores.

We can be quite certain that tomorrow's virtual neighborhoods will mark an important and big use of the Information Marketplace. Remember that bulletin boards and discussion groups were among the first and most popular application babies of time-shared computers and early networks. Today's exchanges, even though they are text-oriented and take place over slow and old-fashioned networks, have resulted in thousands of active groups. And their number is increasing even without the benefit of the better human interfaces and more solid information infrastructures that are coming.

To see where such technical improvements might lead, imagine a virtual neighborhood formed by several amateur musicians, all roughly at the same level of ability, all playing an instrument of their choice, and all interested in the same genre of music, whether it's Gregorian chant or heavy metal. On any given day or night, you as a member of this group would be able to find enough willing virtual neighbors to put together a quick concert for your own pleasure. You would play, each from a different location, sharing the camaraderie of your successes and mistakes and the joy of listening to the common music that you would produce.

In between these virtual performances, you might decide to improve your playing by substituting machines for people. Programs available from entrepreneurs or more advanced musical groups in the Information Marketplace would stand ready to accompany you, simulating several fellow players. You like these simulations because they help you develop your group playing skills while protecting you from the embarrassment of having a whole bunch of people stop playing when you goof.

Such programs have already been constructed in laboratory settings: A decade ago, Professor Barry Vercoe of MIT's Media Lab implemented a successful accompaniment program in which the computer played the piano, sharing the same sheet music with a human violinist. The machine accompanied the human by listening to the violin and learning to adapt to the violinist's speed and individual playing patterns—not an easy task with the highly expressive Romantic music involved. More recently, the people at Coda Music Technologies in Minneapolis have done related work in their Vivace system. Of course, the coordination of players at a distance who cannot "feel" the other musicians next to them through all their senses poses new challenges that will also have to be tested in real settings.

With the Information Marketplace in existence, we can imagine thousands of similar scenarios, representing the world's hobbies and avocations, happening in hundreds of thousands, if not millions, of virtual neighborhoods—all churning out pleasure to countless participants.

Games and Adventures

Community games are the focus of some of today's most popular virtual neighborhoods. As is to be expected, the interactions are mostly text driven and the games mostly adventure. Hordes of noble characters representing the players find themselves in a similar predicament, say, inside a castle looking for gold. The players run into one another, sometimes helping or hindering their mates, subject to the rules of the game and human nature. This is what the multiuser dungeons, or *Muds,* made possible when they first appeared on the Internet. Increasingly, however, Muds have been evolving toward broader virtual neighborhoods where people can hang out and interact on all sorts of topics.

From the vantage point of tomorrow's Information Marketplace, these games will seem lame. With improving bandwidth, computing capabilities, and interfaces, they will certainly expand to faster, more realistic, and more complex actions, incorporating speech, goggles, window walls, and maybe even three dimensions and bodysuits. Yet even though this kind of flashy, action-oriented game is what most people envision when they think of computer games, they are likely to be in the minority, because there will be so many other types of games around that will exploit in the broadest possible way the distributed environment of people and machines.

All of these games will be variations of groupwork software. We will simply perceive them as social games with different names and purposes. Encounter "games" will involve meeting new people in new and different ways. Some may require you to appear disguised through your avatar. They will be the electronic equivalent of the masked balls, where "removing the mask" may be at the option of the wearer or on completion of some goal.

Team games of the we-versus-they variety will be another popular avocation. They will range from adventure and conquest to word games, charades, and the construction of buildings and cities—a bit like building sand castles with your childhood friends. We can be sure that almost all of today's parlor games will have their equivalents in the Information Marketplace.

Card games like poker will fit right in. Twenty Questions might end up exactly as it is played today, with the extra capability of checking the accuracy of the human responses through an online encyclopedia. Discussion-oriented games and groups will move toward real verbal discussions, with gestures and smiles included. When you joined such a group, you would play the snippets recorded over several weeks, hearing and watching an evolving discussion that would seem to have taken place in a much shorter time.

Wagering is another category of game (or, for some, obsession) that will mushroom in the Information Marketplace, because it is rooted in information and the human attraction to quick riches. Pari-mutuel wagering on human contests is another direction these games may take. In jai alai, for example, the odds are determined by the wagering of the people in the house. Imagine what it would be like if all the people on an electronic gaming network could wager on the same contests. Imagine further a nation or the whole globe wagering and determining the odds on issues and events of common concern. In a timid variant of such games, a Web site has already appeared where people can place bets on any prospective assertion they decide to post or find already posted, such as the election to office of a certain politician, the conviction of a criminal undergoing trial, the end of the world by 2010, or the Dow Jones average one year hence. Because there is no monetary gain from this game, much of the pleasure lies in gauging people's interests and opinions through the fictitious wagers they place.

Like the time-shared horserace we discussed, tournaments of all kinds will appear in the Information Marketplace, with numerous kids and adults from different neighborhoods, towns, and nations taking part toward local, regional, and global championships. Games of kinetic skill, mental concentration, and cooperation will lie side by side as the major competitive events. We may even see the rise of a new and more egalitarian Information Marketplace

Olympics where you and I and anyone else will have a chance at competing in something we know how to do well.

No doubt, new avenues to gaming and pleasure will be opened. Ultimately, however, and as in the case of art, it will not be the whiz-bang nature of these games that will make the biggest mark. The real gain will come in the huge numbers of people who will "play" with one another, bridging distances and differences across the globe and getting to know one another as we have done for centuries with games that require physical proximity.

New kinds of pleasurable adventures will emerge that could not have existed before. Consider "Tagalong"—a new kind of explorer machine born in a freshman seminar I teach. Called Hanging with the Hackers, the seminar's goal is inventing fun things. The Tagalong is a robot. It's basically the shape and size of a small, round wastebasket with a briefcase handle on top of its domelike smiley face, without arms or legs, and with a couple of loudspeakers so it can talk to passersby. A built-in microphone records what people say to it, and a small still camera records what they show to it. Equipped with a small computer and a cellular phone, Tagalong sits patiently on a street corner.

If you walk by it, you will hear it call, "Anyone out there?" If you say yes, it comes back with a brief introduction of itself and its goals: It tells you that it is an adventurous robot from the MIT Laboratory for Computer Science touring greater Boston and looking for all kinds of experiences that it will record for others to share. It then asks if you will take it with you and tell it or show it some interesting things. You oblige, pick up the cute little robot, and take it home or to your dorm, where you show it the faces of your roommates, your pet, and a boat model you built. You also tell it what you think of the president and share some of your beliefs and wishes for the world. At some point, Tagalong interrupts you and asks to be plugged in because its batteries are getting low. Once Tagalong is recharged, it asks you if you are by any chance going to Harvard Square. You weren't planning to, but you decide to go anyway just to drop Tagalong on another street corner, which is what it wants you to do. Throughout your encounter, and between encounters, Tagalong has been sending information back to its Web site so that anyone in the world can track where this new Odysseus has been and what adventures it has encountered.

As we have seen, the Information Marketplace is destined to modify and augment our pursuit of pleasure in many different ways. A lot of them will feel natural and will fit nicely into our lives. Others will raise serious questions about our morals and the meaning of human proximity in the new world of information. We'll discuss these important topics at the end of the book.

7
Health

Guardian Angel

Each of us has a "pleasure" side and a "duty" side. We've talked enough about pleasure to get a glimpse of the changes ahead. Let's now shift our attention to our more significant, lifelong pursuits. Chief among them is the care for our health.

The second major explosion of the Information Marketplace, following the entertainment transformation already under way, will take place in health care. In a few short years, medical specialists from around the world will be available whenever you need them. And your health parameters of the moment, along with your life-long medical records, will be available to any distant physician— under your control.

Consider an exciting new software module called "Guardian Angel," being developed by Professor Peter Szolovits and his MIT Laboratory for Computer Science research group. A newborn baby would acquire a module at birth and keep it for life. The module would be distributed, residing partly in the guarded person's portable device, home computer, and personal physician or HMO computer.

When little Barbara Smith is born on October 2, 2005, in Dayton, Ohio, her parents will digitally sign the authorization for her Guardian Angel and the hospital will make the first entries—the minute she first gasped for air, her weight, length, heart rate, fingerprints, and photo. As Barbara matures, the software will keep track of her medical history, step by step. Her childhood bout of measles, her broken left tibia, her irregular early menstruation, her move from a solo practitioner in Dayton to a health maintenance organization in Palo Alto, California, her ten-month use of Prozac in her early twenties, the blood transfusion she needed when her second child was born by cesarean section, the lump in her breast that proved to be a cyst, the diet that helped stave off osteoporosis, and her annual pacemaker tests will be recorded, along with doctors' diagnoses, comments, treatments, and prescriptions.

At this level of functionality, which could be technically deployed today, Barbara's Guardian Angel module will provide a complete medical history of her life, centered about her—the patient—instead of broken into pieces at various doctors' offices, clinics, and pharmacies. It will help a doctor or pharmacist on any given day to retrieve details about her when it becomes necessary. Barbara will control access by setting passwords on the system through her own computer; finally, after decades of bucking an often impenetrable medical establishment, the patient will have some control over her own personal files!

At a more ambitious level, Guardian Angel could perform basic accountability tests, alerting patients and doctors to incomplete or incorrect procedures or behavior. It would give each of us a little more power in the health care process, too.

Imagine, for example, that you are on a ski weekend at Snowmount, outside Salt Lake City.

The powder is fresh and the sky deep blue. Exhilarated, you take a steep downhill run a little faster than you should, catch an edge, and twist your right leg badly. Your knee burns with pain. You worry about the torn cartilage that was fixed with arthroscopic surgery two years ago. Ski Patrol takes you to Dr. Callahan, the on-call leg man at the resort's small infirmary, who's seen worse and has three other patients like you in the next room. You tell him of your prior operation as he eyes your knee, and you give him the access code to your X-ray file back at your home clinic in Minneapolis. Still, you feel concerned, because he seems a bit too casual and know-it-all. After a brief exit he returns, assuring you that you'll need only a splint and some painkillers.

Not convinced, you run a check on Callahan while he's examining the other injured skiers. Pulling your personal computer device from your parka pocket, you press the default button that connects it with your home clinic's computer. The machine responds. You punch in a private code and ask it to display the last time the X-ray file was accessed. The date appears on the device's small screen; the file hasn't been consulted since you sprained your wrist playing touch football last summer. Callahan never looked at your knee X ray. You command the friends you are skiing with to take you to a real hospital and inform Callahan on your way out that you've already entered his name in your insurance company's complaint file. You suspect that he will not be paid for his services and that some form of complaint, after it has been verified, will flow from your insurance agent's accountability file to a few more databases, which soon enough will be accessed by the various sniffer programs that offer advice on local doctors.

Once properly patched up at the University of Utah Medical Center, you collapse for a nap on the couch in your chalet. Two hours later your personal device wakes you with a beep. It's your Guardian Angel reminding you that you have not picked up your anti-inflammatory medication from the nearby pharmacy. You feel your knee throbbing, so you hurry your friends off to the store as you command your Guardian Angel to handle the details of the transaction. It goes on to charge the prescription to your insurance company and notify the pharmacy of the payment.

At an even more ambitious level, Guardian Angel could detect potential conflicts between prescribed drugs, inconsistencies between past and current treatments, and patterns that may forecast potential illnesses. As we know, our future health depends on our present condition, just as our present condition stems from our past habits. A prudent Guardian Angel, programmed with forecasting abilities, can predict how your behavior today will affect your health tomorrow. If you are tall, have a large skeleton with lordosis of the spine, are overweight, and visit the meatloaf table instead of the salad bar, you are increasing your potential for back trouble. This custom forecast directed at you is much more likely to prompt you to change your behavior than some generic advice that you might read in a magazine; it won't be easy to say, "Who, me? I don't fit this pattern," and go on with your self-destructive ways.

A more advanced Guardian Angel could analyze subtle interactions between your behavior, your body chemistry, and your drug intake. A Guardian Angel capable of this work would rise in its usefulness from the level of lifelong bookkeeper to lifelong health advisor. It could inspire you to make changes that would improve your own health and longevity.

The Medical Information Marketplace

A big reason the Information Marketplace will change health care so soon and so profoundly is the pressing need to lower the soaring costs of medicine. An equally big reason is the increase in speed and quality of medical care and procedures provided by the new world of information. Yet another reason is that medical people crave the latest information technologies. Doctors are among the leading users of advanced beepers, videoconferencing, imaging technology, and other trendy gadgets. Already, hospitals, clinics, and large HMOs have begun the big process of merging the Information Age with medicine. They are gearing up systems that will place computers next to every medical professional and provide new software for a host of different medical procedures. The evolution of the medical Information Marketplace is not a theory or a tentative prediction. It is an assured development already under way with increasing vigor in the mid-1990s.

The growing preoccupation with reducing costs and increasing efficiency is good for both providers and receivers of health care. It should not, however, skew our judgment or our attitude about the role of information in health care relative to other factors. Consider this: The annual cost of health care in the United States is well over $1,000 per person, with a life expectancy of seventy-three years, whereas in my native Greece the cost is nearly ten times lower, with a greater life expectancy, seventy-five years. So although the Information Marketplace shows great promise of increasing the quality and reducing the cost and duration of normal medical procedures, we must keep in mind that other factors beyond information technology and money can be even more important in shaping the health of the world's people; reduction of stress and preventive medicine are two leading examples.

Physicians agree that the bulk of our medical problems can be prevented, or at least mitigated, through foreknowledge. It's one thing to have an annual medical exam and another to receive daily information about a condition that needs watching. Because information about an illness (from symptoms to habits) precedes the onset of that illness, "catching" the information means that the illness might be prevented. People should have the choice to accept or reject such prompting information. But in the long run, it would be likely to lead to a healthier population and a more economical health care system than the mend-it-when-it-breaks approach that is still a large part of today's medicine. This is another indirect way in which the Information Marketplace can help improve our health.

Granted that big changes are ahead, what might they be and how might they come about?

At a minimum, the Information Marketplace will make the recording and sharing of billing and treatment summaries more efficient, continuing twenty years of computerization that has already lowered the cost of administrative functions. At the same time, increasingly sophisticated health care providers will strive to use the Information Marketplace to access affordable expert medical talent and knowledge in order to reduce their costs and deliver superior services at lower prices.

An increasingly mobile population will require that health information be shared based on need. An increasingly network-literate population will demand access to its own files, too. Thus, the electronic data available in the Information Marketplace will move on to encompass text summaries of medical histories; then vital numbers like temperature, blood pressure, and pulse rate; then X rays, CAT and MRI scans, ultrasound images, specialized measurements, and the many photos, diagrams, and graphs that currently fill your paper medical file. At first, all this information will be electronically scanned into computers from the paper documents produced by today's medical instruments.

The medical information infrastructure being put in place to make all this possible will initially consist of computer servers in clinics, hospitals, and doctors' offices that hold patient information. There will also be workstations at every nurse's station, testing laboratory, and pharmacy. Electronic communication lines will ensure that information can flow swiftly and reliably between sites. Software, often tailored to each medical center's strengths, will enable the straightforward exchange of text, images, and data.

Because this system will be little more than a speeded up version of what is in place today, it generates little excitement or publicity. Never mind—these "boring" advances will quickly grow as physicians across the globe use the World Wide Web and other network software to exchange text and images conveniently, creating enough of a revolution to be noticed in the quality of your treatment and the cost of your health care.

Meanwhile, new versions of old devices like thermometers, blood-pressure cuffs, and EKG units will appear with the right plugs and software for hooking directly into a clinic's computers— and thus, the global medical Information Marketplace. Medical instrument readings will feed directly into these machines as nurses read them at a patient's bedside. The entire cycle of medical information—acquiring it, shipping it to specialists, analyzing it, archiving it, and accessing it—will become increasingly automated, resulting in greater accuracy and efficiency.

This is already happening in part. Monitoring equipment, some of it in mobile emergency vehicles and some in the residences of home-bound patients, will help people and machines monitor at a distance vital parameters for safety, diagnosis, or immediate action. A widow alone in her small house at night might not require a third-shift, in-home nurse if she connects to her own bedside monitor when she goes to sleep. The machine could send a live feed of her vital signs to a local nursing clinic, where a trained technician watches over it—and data from several other people. The woman's son across town could receive the same feed on his own home monitor. Setup, including local phone charges and the technician's nightly services, might cost $20—a lot less than the $200 a night for an in-home nurse.

The Information Marketplace will also enable doctors to confer at a distance about a specific patient. The medical kiosk in Ruby Creek that we saw earlier is one example. Another example we've also discussed involves advanced operating theaters instrumented with integrated audio and video capabilities. Specialized conference rooms are already being built that will allow doctors worldwide to simultaneously examine the same scan or drug test, even a patient's body. When the lump in Barbara Smith's breast is detected by a scanner in a Palo Alto lab, the size, density, and image would be sent directly to her preferred specialist in San Francisco, who would provide feedback while Barbara waits in the lab's reading room, eliminating the time and cost of a second doctor visit, not to mention a week's worth of anxiety for Barbara and her loved ones.

The computers and communications systems for such advances will be coordinated by middleware modules. Pipe managers will ensure that X rays, CAT and MRI scans, and lowlier graphics and text annotations go where they should go, at the right levels of speed, security, and reliability. Automatization tools will be very handy, because so many medical and administrative aspects of health care are based on standing procedures. The Guardian Angel's forecasts, for example, will be entirely rooted in such procedures. Groupwork modules will be invaluable for doctor-to-doctor and even doctor-to-patient consultations, especially in the delayed, time-stamped mode, because doctors are very busy and are generally unavailable to one another and to us at the same time and place. Security modules will play a key role in protecting medical information from unauthorized use and in handling payments. Hyper-organizing tools will be vital in keeping doctors current with ever-expanding medical knowledge and certified. Medicine will also benefit as a proving ground for novel human interfaces that monitor, help, augment, and even replace some human functions.

In short, the health care portion of the global Information Marketplace will be a veritable orgy of technical sophistication. This fascinating world is already unfolding around us. Doctors at well-equipped hospitals can now share MRI images at a distance. Operating theaters have been built that provide multicamera video broadcasts, which can transmit operations in progress to distant experts. Much more of what we have just described is under way.

Let's move on now to a scenario that illustrates the use of medical expert knowledge. As we go through it, try to guess how far we may be from achieving it.

Automated Specialists

A middle-aged schoolteacher with a known heart condition has been having increased arrhythmia and other heart symptoms. She visits her doctor for help. After an alarming exam, the doctor decides to use the drug digitalis without delay, but he first consults a knowledge-based computer program on digitalis that is available in the Baylor University Medical Information Marketplace. With a single pointing gesture on his computer screen, he sends the patient's history, residing on his own computer, to the distant automated expert. He keys in the dosage he is about to administer and pauses for confirmation.

The program startles him by warning that the intended dosage represents a severe, possibly lethal, overdose. The doctor, now indignant, asks the program to explain its reasoning. After three progressively more detailed explanations driven by his questions, the doctor whispers an embarrassed, "Aha!" Because of an old kidney ailment detailed in the patient's history, which he overlooked, the digitalis excretion by the patient's kidneys would be considerably slower than normal, leaving too much of the toxic substance in the body for too long. The otherwise routine dosage could have put the patient into cardiac arrest. The doctor, no longer resentful, revises the dosage downward but settles at an amount a bit greater than that prescribed by the machine because the patient is a big woman—a fact that the computer program did not take into account. Oh, how human for the human to have the last word! Meanwhile, the patient who could have died under purely human care is saved.

How far in the future is this scenario? Actually, the "digitalis advisor" was developed in the mid-1970s by Professor Szolovits's research group at MIT, with collaborating doctors at New England Medical Center under physician Steve Pauker. Patient experiments took place in Boston and at the Baylor College of Medicine in Houston. A single computer was used. Analysis of fifty cases showed that the dosages recommended by the computer and the doctors were close in all but one case—the woman who could have died.

The digitalis advisor never made headlines, and it was not pursued further primarily because of liability concerns. If you guessed that this example was set in the future, you are like most people who hear this story for the first time. After all, it sounds like something you'd hear in the popular press today about the "future" of information technology. This reminds us how uninformed and out-of-date the hype surrounding the new world of information can be. It also tells us that the incorporation of narrowly focused medical expertise in future systems like Guardian Angel is predictable because it has precursors in earlier research.

Robotic and Augmented-Reality Surgery

The future of health care becomes even more remarkable when we consider the combined progress of several new technologies. The confluence of the burgeoning Information Marketplace with improving robotics will lead to invasive procedures that combine diagnosis and treatment into one step:

> In a hospital operating room on a Tuesday morning, a surgeon makes a three-centimeter abdominal incision in a middle-aged man with a suspected gastrointestinal tumor. She inserts a mini-robot the size of a marble into his intestine and turns to her screen, where she sees all that the robot's tiny camera sees superimposed on a previously recorded MRI scan of the man's abdomen. She carefully manipulates a joystick to control the robot, which is moved by large electromagnets on either side of the patient. Under her control, the little rover slowly navigates the man's intestine, like a miniature submarine. She steers it to the suspect area using the MRI scan as her guide.
>
> Soon, her screen fills up with a crisp, magnified image of the tiny but potentially dangerous growth that the MRI scan portrayed only as a suspicious shadow. She activates a tiny, sharp scoop that emerges from the

rover. With a few delicate maneuvers, she directs the mini-robot to carve out the growth. Retracting the scoop, she navigates the rover out of the patient's body and sends the excised tumor to the lab for a biopsy. The patient goes home, and his anxiety is relieved the same afternoon when the news arrives that the tumor is benign.

Is this bizarre scenario ever likely to happen? Yes, because the various pieces seem to be falling into place. Two people in their early thirties—Janey Pratt, a surgeon at Massachusetts General Hospital and her husband Gill Pratt from MIT—have begun collaborative research on such controllable mini-robots for surgery. Eric Grimson of MIT's Artificial Intelligence Laboratory is already performing research on manual brain surgery guided by MRI scans registered on the patient's skull. This procedure is in fairly regular use at Brigham and Women's Hospital in Boston. Similar research in image-guided neurosurgery is carried out in Alan Colchester's VISLAN project at Guy's Hospital in London and by Philippe Cinquin's group at IMAG in Grenoble, who also use the technique for pedicle screw insertion in spine surgery. The result of these procedures: less invasive surgery; fewer complications, because the surgeon is guided to the right location; no painful waiting for results; and the capability to undertake some surgical cases that were not possible in the past.

Of course, the possibilities are not infinite; technology has many constraints. And some limits on the medical Information Marketplace will come purely from our own very human good sense. Though we as patients should and will allow certain intrusions into our bodies for the sake of a cure, we will refuse others. For the past thirty years, while we've applauded the use of mechanical prostheses for disabled people, it's been possible for us to augment a healthy bicep/tricep pair with hydraulic pistons that would let us have many times the strength of our natural muscles. But we haven't. And as we noted in our discussion of interfaces, we also haven't sanctioned the wiring of our brains directly to a computer. To those who still object to this position as old-fashioned or spiritually driven, I ask a practical question: If we haven't invaded our skin with embedded hydraulic pistons, what will make us invade our brains with embedded computer chips in the future? The answer, I think, is simple: we have respected the sanctity of our bodies and feared the consequences of such meddling.

The Information Marketplace will prompt us to draw other lines in the name of humanity, too. A remotely controlled mini-rover seems only a small step away from remote tele-surgery:

A specialized surgeon places his skilled hands inside two sensitive manipulator gloves that are connected to a computer. Eyeing a high-resolution television monitor in front of him, he picks up a scalpel and begins to operate on his patient—who lies sedated three thousand miles away. The movements of his utensils in open space over an empty table in Seattle direct nimble robotic scalpels perched above a supine patient lying on a real operating table in Atlanta. As the doctor pushes his scalpel down through the air in Seattle, he feels a familiar resistance imparted by his haptic gloves, and the sharp knife in Atlanta descends and pierces the man's groin, exposing a hernia deep inside him.

A likely scenario? Maybe some distant day, but don't hold your breath! The basic approach has been demonstrated in a rudimentary form by ARPA, with a surgeon suturing organs in a dead animal from a small distance. In a more realistic tele-surgery setup, the reliability of complex systems that combine computer hardware, software, and communication links would be far too shaky to be trusted with people's lives. Even NASA's computers, designed for maximum reliability, are known to fail every few launches. So until the reliability of these systems becomes rock solid in some future decade, I suspect that, like me, you would rather be the doctor than the patient in any such tele-surgery scenario. No matter how far technology advances, we should—and undoubtedly will—always feel free to exercise our human prerogative by saying "No."

8
Learning

Experiences and Prospects

Like health, education is a lifelong pursuit that the Information Marketplace will surely affect in many ways. But will these changes actually improve learning?

Ever since computers were deployed in earnest in the 1960s, researchers have been trying to use information technology to improve learning. Today, numerous such experiments are under way on the Internet. In one project, grade-school students around the world are collecting measurements of their local environment. The information is entered into a common database, building up a global environmental model shared by all the contributors. Besides learning about their surroundings, the children are learning how to communicate and cooperate with children of different cultures.

This exciting project was initiated in 1995 at the G7 conference in Brussels, an annual meeting among the world's seven wealthiest nations, which that year focused on information and society. It is an application of groupwork tools to learning, which may well be the pivotal technology to yield long-awaited breakthroughs in education. Learning has always been a people-to-people experience, so it stands to reason that a computer-mediated people-to-people process may be just what is needed.

And yet, except for a few cases, there is little proof that such exercises actually improve learning. Do they help retention? Build complex ideas from simple ones? Improve problem-solving skills? Provide necessary perspective? We don't know . . . yet. Even though hundreds of rich and promising experiments have been tried for twenty years, the jury is still out. It is unclear whether computer and communications technologies help the learning process in a fundamental way. We have certainly discovered exciting ways of using information technology in learning. But we must be conservative when it comes to the education of our children. It is simply not enough—and may be damaging—to gaze with wonder at a novel technological approach and declare it educationally effective just because it is exciting.

Here's a scenario with lots of promise:

It is Monday morning and a casually dressed Seymour Papert, an MIT professor of mathematics and disciple of the great psychologist Piaget, is standing among fifty researchers unveiling a grand new programming language called Logo. With it a child can enter simple instructions on a computer that direct a mechanical turtle to move around the floor and draw pictures using a pen that drops and retracts from its belly. I take my turn and type the following:

Pen Down
Forward 10
Right 120
Forward 10
Right 120
Forward 10
Right 120
Pen Up

The turtle obediently moves forward by 10 units, turns right by 120 degrees, and repeats this ritual three times, thus drawing a triangle with three equal sides on the paper on the floor. I give this drawing program a name by typing "Triangle."

We researchers are so excited by Logo's simplicity and power that we head off to show it to a fourth-grade class at an elementary school in neighboring Cambridge. Of course, the boys and girls catch on immediately. A Korean boy types "Triangle" then orders the turtle to repeat the new program thirty-six times, turning 10 degrees each time:

Repeat 36
Triangle
Right 10

The turtle dutifully creates a brilliantly symmetrical figure that no child would have had the patience or skill to draw. The children vie for the controls all morning long. Soon they teach the turtle to write letters and form sentences by randomly grouping words from a list they supply: "Mary loves John. Anna and George have cooties." They explode in elation as only children can. With this fun "game" they are learning geometry and English—not to mention how to program a computer—and loving it all the way.

Like the digitalis story, this sounds like something you'd hear in today's press about the wonderful potential technology has for improving education in the future. But it occurred in the early 1970s—another (and I promise the last tricky) reminder of how out-of-date some predictions can be. The initial work on Logo was done at BBN by Seymour Papert, Cynthia Solomon, and Wallace Feurzeig. Related research in computers and education has since been carried out by many people, including Andy di Sessa at the Graduate School of Education at the University of California at Berkeley.

Logo didn't cause a revolution in educational processes, although it certainly became a useful tool after Papert and his colleague Hal Abelson perfected it. Today Logo is used in many schools. The early demonstrations convinced most of us researchers that Papert had unlocked a whole new world of "learner-driven learning." Educators had long known that boring memorization and drill are okay if they are carried out toward a desirable goal, like learning how to drive a car, but not as exercises unto themselves. Children—and the children we all are—prefer to learn through the excitement of discovery and participation. However, school kids seem to learn just as well in less technologically sophisticated ways. We have since come to realize that technology alone, no matter how futuristic or exciting, doesn't automatically improve the learning process.

The lack of a major breakthrough in a mere two decades shouldn't be construed as a license to perpetuate tired teaching methods and avoid innovation. And yet "the old way" is preferable to blind adoption of information technology in schools, *on an irresponsibly widespread scale,* based on the desire to appear modern and the naive assumption that if the technology is deployed en masse, surely a thousand educational flowers will bloom. Well-meaning politicians have been among the worst culprits, calling for wholesale changes because the topic seems important and fashionable.

So what should we do? We should use what we know works, and we should experiment with new ideas actively and intensely on smaller numbers of students—especially because promising approaches on the World Wide Web are mounting up for trial. I will shortly explain why I believe that the Information Marketplace is destined to provide new approaches that will truly improve learning. Until this happens, and even to help it happen, we need to continually examine what succeeds and fails, and why. And we should do so *before* we deploy any technical approach on a grand scale.

This experimentation will require money—and a flexible attitude toward change. In the United States, we citizens agree with one another and through our politicians' pronouncements that nothing is more important than the education of our children. But when the time comes to reach for our purses, we refuse and vote to cut local taxes, thereby squeezing school budgets. In Europe and Asia, where school money is centrally supplied, there is a different barrier: a greater resistance to change. Neither money nor flexibility can solve the problem; both are needed, together with some good luck, if we are to innovate.

Let's now take a quick look at the different ways we might use the Information Marketplace in education. Some of the approaches have been tried, in which case I'll report on the outcome. Others are yet to be tested; in these cases we'll assess their chances to improve learning. As we go along, keep in mind that the techniques apply to all educational settings: grade school, high school, the university, continuing education, career development, even a how-to guide for home repair. We'll start with simple approaches and crank up the complexity and sophistication as we continue.

Homework

At the simplest level, teachers can use computers to hand out homework assignments, receive students' responses, and return graded work. Some teachers have been doing this for twenty-five years; they've found it offers some logistical benefits but that there is no improvement to education. The Internet has brought new twists. Schools are beginning to put homework assignments on their home pages that both students and parents can access. No longer will young Nicholas be able to claim that "the teacher didn't give us any homework" or "I don't remember my assignment." A commercial service called Homework Helper has been started in Philadelphia by the Infonautics company, which lets students use plain English to access an extensive online library of thousands of magazines, newspapers, reference books, photo archives, and more. Now, Nicholas won't even be able to say, "I don't have the right reference to do my homework."

More trendy are specialized "knowledge hyperdocuments"—software tools that organize knowledge for student use. They include today's hypermedia—the so-called *edutainment* CD-ROMs for personal computers that present numerous snippets of text, pictures, video, and sound. If our boy Nicholas is reading about Christopher Columbus on his computer, sees the word *Niña* highlighted on his screen, and clicks on it with his mouse, he will be transported to a drawing of the ship. If he clicks on the captain standing at the bow, a portrait of Columbus will fill the screen. Reading the short story that appears with the picture, Nicholas might click on the name *Queen Isabella* and be transported to the Spanish royal castle where she gave Columbus his funds. And it can go on and on for as long as Nicholas wants to explore, skipping between vignettes chronologically, geographically, or topically.

Hyperdocuments like this represent a significant departure from the linear organization of knowledge used for centuries in books. But is this kind of branching educationally more effective? The point-and-click excursions are driven by the learner, rather than the teacher, and we know that this is good for discovery and motivation. But will Nicholas remember the ships' names, or understand the debates Columbus had with Isabella, any better than if he had spent the same amount of time reading a few pages in his old history book or listening to an explanation from his teacher? The advantages of empowering students to control their own learning may be offset by loss of the guidance of an experienced teacher. In most cases, combining both approaches may involve more time and effort than the student or teacher can afford. After the hype subsides, it may well be that hyperdocuments will have found their niche in learning as very useful albeit overglorified indexes.

"Analysis tools" are computer programs that can help a student probe for underlying causes of what is happening in a specialized area of knowledge. They do best in quantitative disciplines like math, physics, and engineering. One program, for example, could determine the voltages and currents in a simple electrical circuit designed by Nicholas (now in high school). Educationally, the approach is not very enlightening; from Nicholas's perspective, the process is similar to having a stern teacher who declares, "This is how it works," without further explanation. However, coupled with a good human teacher or the invitation to "read" and understand the program or spoken dialog that provides explanations, analysis tools can help students increase their understanding and sharpen their intuition.

Simulators

Up one step are "simulators"—computer-driven machines that present realistic situations, popular for the training of commercial and military pilots. Our young man Nicholas, now an Air Force cadet, climbs into a mock F-16 cockpit and is presented with convincing video of views out the windshield and readings on cockpit displays. As he operates the "plane's" controls, high-fidelity speakers transmit flight sounds while mechanical pistons vibrate and tilt the cockpit, creating a full-body experience of turbulence and g-forces. The simulation is so effective that commercial airlines use these contraptions to certify pilots on planes the pilots have never flown. For Nicholas, the stress in the mock environment is so real that he exits the simulator sweating in fear.

Simulators are ideal for learning kinetic skills, and early versions have already proven themselves. Many American tank commanders in the Persian Gulf war received a good part of their training in simulators before heading for the desert. With special joysticks and other gadgets, we common citizens will use simulators extensively in the Information Marketplace to help us learn how to fix an appliance, ride a bike, ski, sail, or drive a car before we try the real thing, leading to more confidence and greater safety. In the professions, too, simulators will go a long way toward teaching kinetic and manipulation skills: surgical simulators for all sorts of medical specialties are likely to become a big hit. Servers offering such simulation services will indeed grow in the Information Marketplace and prove very useful to human training.

Among the different ways of using information technology in education, simulation is a clear winner that truly improves mechanistic and kinetic training.

Can simulators cross from mechanical to qualitative situations? Might we ever develop simulators that help executives learn how to handle a management crisis? Would it be possible for Nicholas, having completed his military service and gone on to a career at Boeing Aircraft, to walk into an enclosed "confrontation simulation chamber" the size of a small cubicle and face an angry leader of the electrical workers' union? Would Nicholas's adrenaline pump just as it did in the rollicking computerized cockpit while he stands in tense face-to-face debate with the union rep?

The answer depends on how well computers will be able to simulate qualitative human behavior, as opposed to the quantifiable pitch, roll, and yaw of an airplane. This might be possible only if the dialog is limited to a narrow domain. It is technically feasible to record a few hundred video sequences depicting the behavior of a disgruntled union rep in response to a small number of different statements a manager might make and weave them into an interactive movie that unfolds seamlessly depending on what the manager says. As we read earlier, there will soon be adequate speech-understanding technology to track Nicholas's statements if the discussion is kept to a narrow topic—say, wages.

Might this approach be extended to showing people how to cope with personal conflict? For years we have taken great pains to teach children mathematics, language, history, and science but have spent precious little formal effort on the important aspects of human relationships. As a teenager of the future, could Nicholas walk into a friendship, marriage, or divorce simulator and be wiser for it? Or learn how to face down a peer on the street corner who tries to persuade him to sniff cocaine?

Nicholas steps into the small, dark simulator. A life-size video begins before him. A tall teenager wearing a slick jacket approaches him from a playground basketball court as city traffic goes by in the background. "Come on, Nick, try it," the kid Jack says, flashing an encouraging smile and stretching out his hand toward Nicholas. "It's free, and it's awesome." "Well, uh, I don't know," Nicholas stammers, intimidated. "Come on, you want to be cool, don't you?" Jack says. "Well, yeah, but . . . uh."

Nicholas didn't do very well. Take Two:

"Come on, Nick, try it," Jack says. "Well, my mom says I shouldn't take drugs," Nicholas replies. "Yeah, well, moms are always telling us what to do, ya' know?" "Well, I don't want to get in trouble . . ." Nicholas finishes, torn between parental obedience and peer pressure.

Better, but not best. Take Three:

"Come on, Nick, try it," Jack says. "Forget it," Nicholas says, having gotten up the courage. "If it's so cool, what are you doing hanging around here by yourself?" Nick walks away. Jack shrugs and walks away too.

Now Nicholas has got it. Without the benefits of this preparation, chances are he might not have resisted a real Jack on a real street corner three afternoons in a row. The practice—and privacy—afforded by simulators may be important in encouraging learning about such nontraditional but nonetheless essential educational topics.

There are technical limits to simulation. We're a long way from being able to simulate Jack's acting and talking with a wide range of behavior. As we have seen, the computer can't understand human speech in broad discussion domains. As MIT's Marvin Minsky, cofounder of the artificial intelligence field, has noted, we don't know how to program computers to exhibit common sense. This severely limits the range and quality of Jack's responses. Even if this huge problem were overcome, a computer cannot yet harmonize faithfully the lip synchronization, nods and gestures, and voice inflections needed for Jack to respond to any random input in a believable way. And if Jack's visage isn't believable, Nicholas won't learn his lesson. A few behavior simulators might impress us with their usefulness in the coming decade in highly specialized situations, but more versatile systems are still a long way off.

This kind of learning is not restricted to simulation. It could take place today with no technical difficulty using a real teacher or an older friend whose avatar is Jack. Jack would encounter Nicholas or Nicholas's avatar in a Mud (multiuser dungeon) where they could have exactly the same conversation, or an even better one because humans are involved at both ends. Although this approach costs more because it involves additional people, it may be a realistic possibility; many more people might volunteer their services for these teacher or helper roles from all kinds of locations and different groups (like retirees, past victims, and home-bound individuals).

Design and Creativity

Sometimes we learn best by constructing rather than taking apart something we are striving to understand. "Synthesis tools" or "design tools" can help us learn through the design of real and virtual objects. One design tool might take your sketch of a restaurant table and render it so you can see how your decisions affect the three-dimensional look and cost of the table. An alternative aid might suggest "similar" designs that have proven worthy in similar situations. Invention Machine Corporation, with U.S. headquarters and employees in Russia, offers a software product that gives helpful and relevant suggestions to designers of electromechanical devices during the creative and construction phases of design.

I tested that system with the same design challenge I give my freshmen in our creative design seminar: Design a round table for restaurant use that can gradually change diameter from three to six feet so that it can adjust to the number of people that sit around it. The system did match the variable diameter request with mechanisms that expand under heat and with the iris mechanism used in camera lenses. As it proposed these solutions, it also produced explanations and drawings of how they work. Creativity seems to involve an interplay between the generation of wild ideas and the somber scrutiny of assessing the utility of these ideas. There is plenty of room in this process for people and computers to work in concert, as the above example shows.

After years of technical management, Nicholas might find pleasant diversion in writing poetry for his wife and daughter. A poetry aid would help him with synonyms, meter, and rhythm, as well as traditional and abstract styles; guidelines would pop up on his computer screen during the creative process. A music composition program could present harmonic analysis and progressions, plus a large library of interesting musical themes, riffs, and percussion patterns. Helpful software applications like these will be technically feasible within the coming decade, making computer-aided creativity as accepted as computer-aided design in many different creative endeavors.

For simpler synthesis tasks, like writing a form letter, constructing a résumé, or designing a device based on routine procedures, programs available today can do essentially all of the work. But that does not help learning or creativity any more than an overzealous parent who solves a student's problem by doing it for him. These turn-the-crank synthesis tools are not very useful for learning and should be used primarily as subordinate aids that support other, better learning tools and our own creative thinking by simply getting the necessary routine work done and out of the way.

Automated Tutors and Masters

A natural extension of synthesis tools are "automated tutors," which customize their contribution to suit an individual's needs. As they engage you with questions and interaction, usually while you try to construct something, they build within themselves a "map" of your strengths and weaknesses. They can then exercise your weak spots by tailoring problems to strengthen them.

Automated tutors are hard to implement, because they must be intelligent to be effective. Yet technically simple versions of such programs can offer substantial tutorial and psychological advantages. Illiterate adults often fail to seek help because they are embarrassed to appear before another adult and admit they cannot read or reluctant to struggle with something the average child can do. But an illiterate forty-year-old man might be willing to lock himself in a room with a computer for an hour each night, learning to match words he sees and hears with pictures of these words on the computer screen, and emerge a month later, able to surprise his kids by reading them stories. Beverly Hemmings, a master of science MIT graduate who dedicated her skills to build such tutors for inner-city families, had several such successes. Programs like her Touch 'n' Tutor we just described and the literacy tutor of chapter 3 have a bright future because they work and because they address an increasingly pressing worldwide need. In tomorrow's world of greater information, attaining universal literacy is not only a moral obligation of society but also a means to greater productivity and improved quality of life. Mark the literacy aids as another sure winner.

An "online teaching assistant" is another type of simplified tutor; it holds "canned" answers to frequently asked questions in a given discipline. A student types in a question, and the system answers. If the program doesn't find the question in its list, it routes it to an on-call human teaching assistant who enters the answer from a remote location. This satisfies the learner and enriches the response library of the online program for future student questions.

Between 1984 and 1991, MIT tested a few automated tutors in Project Athena, which dealt with several educational experiments using computers. One program was an online teaching assistant for math and physics problems. Another program guided a student in analyzing mechanical structures; it suggested options when the learner got stuck or had gone down a blind alley. In yet another Project Athena experiment, pioneered by Janet Murray and Gilbert Furstenberg, students learned French through an adventure game set in Paris. The student had to help a French journalist find an apartment. She could only communicate in French. Virtual people, including the journalists' friends, relatives, and Realtors, responded to her questions and posed questions of their own.

Unexpected events also happened: a plumber might walk into the journalist's apartment and demand to be paid. The element of surprise and the realistic setting reinforced the student's learning.

The French learning adventure proved effective, but it was difficult and expensive to develop. Like the other major experiments, it devoured a million dollars or more. These advanced learning tools are likely to evolve slowly in the Information Marketplace and only in selected areas of learning. Their steady and patient development is well worthwhile and should be encouraged.

Many other computer-assisted educational experiments have been carried out in projects like Athena at Brown University and CMU and more recently at the Institute for Learning Sciences at Northwestern University. At the Rensselaer Polytechnic Institute, there are no longer physics lectures. Students work in a computer-based studio interacting with one another and with software tools.

The most elevated future application of information technology to learning is "automated apprenticeship." Let's say Nicholas responds to his midlife crisis by deciding to augment the technical design skills he learned in the Air Force and at Boeing and become an architect. At night, at home, after learning the basics, he fashions various designs on his computer under the watchful eyes of Frank Lloyd Wright, a famous twentieth-century U.S. architect, and Ictinus and Callicrates, who designed the Parthenon in the fifth century B.C. These legendary experts, whose virtual presences are synthesized by the computer, point out the strengths and deficiencies of Nicholas's designs, recommending improvements and alternatives—just as the Japanese architect did for Mary's theater designs.

If they are developed at all, automated apprenticeships are still very far into the future, because they require computers to exhibit nearly humanlike understanding. Although we still understand little about human cognition, researchers are slowly gaining on programming narrow models of reasoning about specialized situations. If (and when) with some good luck automated apprenticeships were to move forward, it would behoove us to encourage the grand masters of today and tomorrow to leave behind a living legacy of their expertise by explaining to suitable repositories their approaches, reasoning, and preferred methods for critiquing students' works. The memory and the legacy of revered historical figures would then assume new dimensions for mankind through their continued presence in our midst—virtual immortality! And if it's good enough for the heroes, why not for the rest of us? Might it not be interesting and even heartwarming, in some future century, to talk with your distant ancestors, even if the conversation is limited?

Staying with high-risk predictions, it may also turn out that computers will open up a new form of knowledge. Seymour Papert and Marvin Minsky of MIT were among the first to recognize this possibility. Alan Kay, who with others pioneered the personal computer at Xerox Parc in the 1970s, has expounded on what the new knowledge does: it can codify and describe events that neither spoken nor scientific language can. Computer programs can capture and convey to people and machines the complex, dynamic, and awesome processes that change the weather, describe a flame, or cause biological growth better than any text or diagram or syllogism. This *new* form of knowledge should open up exciting doors for learning about the world around us.

The Changes Ahead

In the short term, the Information Marketplace will improve learning with the more successful approaches we have just discussed. It may not improve learning across the board as rapidly as it does health care because of the lack of funding, social flexibility, and inconclusive research to date. Nevertheless, the longer-term possibilities are very promising. Why do I think so, in light of our many years of mixed results? The first Industrial Revolution affected education indirectly, in the sense that better-fed students made better learners. The second Industrial Revolution continued to help indirectly through better transportation of students, better heating and lighting in schools, and a population wealthy enough to send its children to school instead of work.

The new world of information breaks away from this indirect pattern of help. It is *directly* linked to the nuts and bolts of education through the acquisition, organization, and transmission of information and the simulation of processes representing knowledge and through the use of approaches like e-mail and groupwork that mediate teacher-learner and learner-learner exchanges. As such, it's the first major socioeconomic revolution in history that offers technologies directly involved in the learning process. It therefore has a good chance (though by no means a certainty) of generating breakthroughs that could not have resulted from the more indirectly linked technologies of the two industrial revolutions.

The Information Marketplace will change the role of schools, universities, and the educational community. One of the more obvious effects will be the simultaneous expansion of the student market for schools and the school market for students. Why study at the local school, training center, or university if you can attend at a distance the best school for your particular interests? This question is causing a lot of confusion and even some hasty actions and declarations about distant learning as schools and universities scramble to take advantage of an information-rich world. Let's look at it more carefully.

In some situations learning at a distance will make a lot of sense. If the alternative is no school, then a virtual school will be better. For years the children of mariners have taken their schooling via correspondence because they don't stand still long enough to go to a school of bricks and mortar. For updating highly specialized skills that emphasize factual knowledge, perhaps of the medical variety, it will also make sense to go to a virtual school that fulfills that need. Already there are Web courses that help physicians upgrade their expert knowledge and become certified in certain specialties.

For the bulk of education, however, such distant learning approaches don't quite work as well as the traditional settings. Education is much more than the transfer of knowledge from teachers to learners. As an educator myself, I can say firsthand that lighting the fire of learning in the hearts of students, providing role models, and building student-teacher bonds are the most critical factors for successful learning. These cardinal necessities will not be imparted by information technology. So even when the jury finally concludes, as I suspect it will, that the Information Marketplace can radically improve learning, teachers' dedication and ability will still be the most important educational tool.

Another vital part of the educational process is joining and identifying with a community: the opportunity to become motivated by role models of fellow students and teachers and ultimately the experience of certain primal forces that we'll discuss at the end of the book. Even schools that offer classes at a distance, will find they need to offer a real-life term every one or two virtual terms—much the way a company's sales team has to come to the home office and meet periodically with their fellow employees. What's more, schools with tough admissions standards will want to control the quality of their student and faculty communities. Just because connectivity is technically possible, that doesn't mean they have to open themselves up to just anybody. We will no doubt see alliances between institutions of comparable quality and complementary offerings for joint-degree programs and synergistic research efforts. Employers too may become more directly involved with their feeder institutions, to get both early notice about promising students and a chance to tout to students their company's advantages. Then again, until the new approaches have proven themselves, employers may have reservations about choosing virtually educated people over those who have been educated at schools of brick and mortar.

Other changes are more certain. Libraries will remain the custodians of physical educational materials, notably books. But they will also become managers of the information links to other knowledge sites, with the important proviso that they, the libraries, control the quality of these virtual bookshelves, deciding which knowledge residing at other institutions should be targeted by the pointers and hyper-organizers of the local library. The new librarians will actively ensure the presence of only those virtual links that preserve a quality and currency of shared knowledge deemed necessary and complementary by their institution. Effective management of these shared knowledge pointers will be critical to the quality of tomorrow's educational institutions, especially because students and faculty will also have access to their own huge arsenal of distant knowledge links.

This discussion of libraries brings up another exciting possibility: the formation of a grand "world heritage library" made possible by interconnecting in a mutually agreed-upon, uniform way all the libraries of the world. Each nation would supply in electronic form their contributions to literature, including rare and out-of-print volumes. To us users it would all look like one uniform library stocking all 100 million books, documents, and other creations of our human heritage. This dream is being pursued by Raj Reddy, Carnegie Mellon's dean of Computer Science, and others worldwide who deserve our support. We can't even begin to imagine the great benefits to learning, to the appreciation of great literature hence to improvements in the written word, and the enjoyment of knowledge that could flow out of such an incredible collection, available to everybody, everywhere, at any time.

The picture that emerges from this discussion is one of a robust Information Marketplace poised to improve education by augmenting and enhancing rather than replacing the physically proximate ways of teaching and learning. In learning, as in health care, the Information Marketplace will produce visible changes that stem from its novel capability to bridge distance, ship around sensory information, and promote group interactivity.

How profound will these changes be for our noble quests of health and learning? In medicine, we have seen that we can expect sizable practical improvements to the quality and cost of health care. In education we understand where computers have and have not been helpful to date, and we are very hopeful, if not convinced, that a breakthrough is in the making. All in all, the potential impact of the Information Marketplace on our "duty" side seems just as ubiquitous and deep as it promises to be for our lighter, more pleasurable activities.

9
Business and Organizations

Group Power

We began this part of the book by focusing on how the Information Marketplace is likely to change our daily lives. We then moved to our pursuit of pleasure and the important lifelong achievement of good health and education. Now we widen our scope from individuals to organized groups of individuals, continuing to ask how the new world of information might affect the groups to which we all belong by choice, necessity, and simple geography.

Companies, churches, universities, and armies are all human organizations. You and I, the participants in these groups, are bound together by the forces of a common purpose and the belief that by acting together we can fulfill our goals better than we can by acting alone. The methods organizations use to pursue their missions involve a great deal of information, from the routine paperwork their office workers seem to thrive on—correspondence, inventory and manufacturing controls, payroll, orders, invoices, accounting, advertising, patents, contracts, and more—to actions such as placing phone calls, holding meetings and corridor discussions, exchanging meaningful glances, and engaging in much serious thinking and planning.

We'll begin with the organizations that are part of our everyday lives—manufacturers, retailers, service providers—and extend our view to social institutions. We'll then make some observations about the most important changes (and nonchanges) that we'll see across the board, from new structures like expert centers and work centers, to issues such as accountability, the spread of egalitarianism, and shifts in responsibility. In chapter 10 we'll address the largest organizations of all: governments.

Business Changes

Too often we hear catchy phrases about how the Information Age will help "manage a company's knowledge assets," "enable the instant organization," or "build network alliances." Slogans like these may be impressive, but they generally offer no clue as to how the Information Marketplace will affect the way we do business. Executives seeking to understand what aspects of their business are likely to change as a result of the Information Age are better off ignoring the slogans and carrying out a simpler, more direct examination: They should first unearth where and how they use information (noun and verb, human and machine) in their own specific business both internally and in their external dealings. They should then assess if and how the new capabilities and approaches made possible by the Information Marketplace can help them use this information more effectively.

To understand how big such changes may be, let's ask how much of a nation's economy deals intensively with information. We can arrive at this percentage by subtracting from the overall economy everything that involves physical and other work not directly driven by information, like transportation, agriculture, and physical services such as restaurants and gas stations. Of course, even in these sectors we must add back their many information-intensive activities such as ordering, accounting, data processing, advertising, negotiating, contracting, selling, shipping, monitoring, and invoicing, as well as managing people, writing letters and memos, making phone calls, faxing, and copying—the grand fabric of office work.

A financial services firm's activities might rely nearly 100 percent on information, whereas a restaurant's dependence might be 5–30 percent. Across the entire U.S. economy, 58 percent of the total workforce (including government) deals with office work. It is also estimated that 60 percent of the U.S. GNP deals intensively with information. Across the industrial nations of the world, the situation is not much different, with the ratio closer to one half. We will discuss this portion of the economy—the "Information Economy," as we'll call it—later on. For now we may safely conclude that *on average, about half of every business and organization contributing to the industrial world economy could be affected by the Information Marketplace.* Let's examine a few of the most notable transformations.

Electronic Commerce

When people speak of electronic commerce, they often confuse the two distinct flavors that it will have.

By far the larger of the two involves handling the information needed to trade physical goods. Whether it has to do with the flow of natural gas along a transcontinental pipeline or the purchase of shoes at World Shop, this "indirect electronic commerce," as we'll call it, handles the advertising, searching, selling, contracting, settling, and other such information-related functions, though the actual products or services are physical goods shipped on traditional transportation systems. In 1997, indirect electronic commerce accounted for over $1 billion and involved the purchase and sale of books, computers, stereo sets, groceries, and cars. Indirect electronic commerce will flourish and result in substantial gains for buyers and sellers because so much of what goes on between them is information.

By contrast, "direct electronic commerce" involves goods that are themselves information, shipped directly through the Information Marketplace. These goods include software, manuals and books printed over the Net, photographs, X rays, medical records, music, movies, travel guides, news, stock prices, money, procedures, forms, educational materials, advice, and all the information we have talked about. Better output devices like glossy color printers, high-fidelity speakers, large displays, virtual reality systems, bodynets, and a host of specialized devices will make this kind of electronic commerce even more attractive. In 1997, direct electronic commerce was beginning to rev up and involved mostly human work like the services of writers and graphic artists. But because the speed-up in delivery satisfies the human need for instant gratification, direct electronic commerce is likely to become an important component of the Information Marketplace.

As with other activities in the Information Marketplace, a good part of the success of electronic commerce will hinge on the degree of sharing and standardization that buyers and sellers can agree on—the automatization conventions we discussed in chapter 4. These tools that will form the new "language of commerce" will evolve within large companies and business sectors as companies, associations, and interest groups agree on what they will do together and how they will do it. As we have seen, one simple yet powerful tool is an e-form that automates or semi-automates searching, negotiating, ordering, contracting, and billing and can also reduce the linguistic barriers in international commerce.

> A central European wholesale buyer of fruit issues an e-form requesting bids for 200 tons of oranges of a specified quality and size, deliverable in one day. His e-form is instantly translated into Greek, Spanish, and Italian (an easy task because of the preagreed meaning of the entries) as it appears before the sellers in the corresponding Mediterranean countries. The sellers may review the order, while their machines respond almost instantly, and a few minutes later the deal is closed!

Here then is the first recipe for organizations aspiring to use the Information Marketplace for electronic commerce: Get together with your peers and common-interest groups and professional associations, even your competitors, and develop simple e-forms that you all agree will save everyone time and money through automatization. This has begun happening predominantly for financial and easily quantifiable data. The ANSI x.12 and the U.N.'s Edifact standards already span several diverse business areas, while large companies, like Ford, are using the Web experimentally to link to their customers and suppliers. In following this course, businesses and associations should ensure that what they are doing is aimed squarely at automatization rather than the narrower objectives of some EDI conventions that are restricted to financial interchanges.

A huge part of both direct and indirect electronic commerce will involve human work. Businesses aspiring to exploit the Information Marketplace should explore how they can use other types of middleware. Groupwork modules can link specialists across space and time. British Petroleum, for example, has had quite a bit of success in enabling its experts around the globe to jointly solve problems as they arise. It is standard practice for oil companies to maintain centralized "firefighter teams" that can be flown to various trouble spots—oil rigs or tankers or refineries. But British Petroleum now brings the expertise of its people located in different countries to bear on some of the same problems without having to fly many, if any, people to the site, reducing delays and costs. In addition to problem solving, B.P. uses this technology for special projects, for their top two hundred managers worldwide, and for links with selected suppliers. They are already beginning to solve problems faster, reap the benefits of deploying the resources that match the problem, and develop viable options at reduced cost.

Company strategists should ask themselves what business functions middleware can improve. Can e-mail be used effectively to deal with customers, suppliers, and competitors? Can certain functions be automated over the information infrastructure? Can groupwork help employees operate across departments? Can people be hired at a distance to process information remotely at lower cost or higher quality? Can frequently changing information be hyper-organized to make it more useful? Which aspects of the business must be secure, and which can be open in the Information Marketplace? Can information-related processes at different sites become integrated for smoother operation? Can the information infrastructure improve advertising, marketing, sales, customization of products and services, manufacturing, and cross-departmental management? Tackling questions like these is the balance of our recipe for organizations and entrepreneurs seeking to exploit the Information Marketplace.

We can envision how electronic commerce might handle these activities by imagining how you might buy a car in the Information Marketplace in a decade or so. The first thing you would notice is that the car manufacturers, with higher-quality information about their customers, will better target their marketing, focusing predominantly on consumers most likely to buy their products and services. So you will probably get less junk mail about cars you don't want. You may also use reverse advertising to declare the kind of car and features you desire and let competitors who can fulfill your specs come to you. If you are tall, you might put out the following reverse ad: "I want a four-door sedan with the greatest driver headroom for under $20,000." Makers of cars with generous headroom will be anxious to deal with you. Makers of cars that do not have this feature won't waste their money and your time trying to attract you.

Another change you will notice will be greater customization of the product . . . before the sale! You will play with a seductive design-your-car kit that lets you dictate the various options you want. You will also engage in what-if simulations, telling an online simulator of the car to perform in mountainous terrains under a heavy load or in heavy city traffic on a hot August afternoon. Eager sellers would also make sure that you see a picture of yourself driving your customized dream car—achieved by blending your image with the vehicle's.

To help you evaluate what car to buy, you might also visit a new breed of advisors, who will offer their recommendations for a fee, answer your specific questions, and give their assessment of the product-buyer match based on your interests, which you would provide through your auto-profile e-form. For another fee, your evaluation would also involve finding out how many people have bought this kind of vehicle in the past, the range of prices they paid, and a summary of their experiences and comments.

In the end, your negotiation with the dealer on a final price would be much less secretive, less intimidating, and more factual because the total range of sales prices for the car would be well known to you. The closing process itself would be more streamlined, too, aided by nearly automatic contracting, registering, and financing.

Such tailoring over a broad range of customizable products could have disadvantages—complex pricing schemes, difficulty in comparing prices among dissimilar goods, and price fluctuations. We must also not forget that there is value in standard goods—they require no effort on our part to customize them to our individual needs, and they may be cheaper. So although many goods, especially expensive products like cars, will become customized using the Information Marketplace, we will still continue to need and use standard goods.

Making Things

The sale we just described marks the beginning of another important process. A good part of the final assembly of your car would take place near you, perhaps in a combination revamped dealership and regional "finishing" plant. Once your order was in and you had made your electronic down payment, the dealership would ship a multitude of instructions to the manufacturing enterprise. Sophisticated automatization programs designed by the company would "explode" your dream car into the countless assemblies and subassemblies that it comprises. The global trend toward minimizing inventory will ensure that much of your car will be manufactured in different locations within a few days on either side of the day you placed your order. Programs will try to anticipate and pre-manufacture long-lead items based on statistical purchasing trends so that you don't have to wait long for your car. All these pieces will be rapidly combined into major assemblies that are shipped to the "dealership" for final assembly.

For these changes to materialize, however, car manufacturing procedures would have to evolve well beyond their current form, which is still largely rooted in Henry Ford's practices of mass production. Carmakers would have to shed the hundreds of huge robots on the factory floor, which take hours to perform one or two thousand spot-welding operations on each auto body. Instead, carmakers might fashion a new breed of composite materials and novel fasteners so that your car's body would be assembled in fifteen minutes by four moderately skilled workers who would snap together perhaps sixteen prefabricated panel assemblies and carry out a few other necessary operations. This is similar to the way people today build computers by plugging together well-defined subassemblies.

Regardless of whether such manufacturing innovations happen or more conservative practices prevail, the carmaker would assemble each car to order and would know at all times the state of production of its various orders. The links between sales, management, and the factory floors will be much tighter, made possible by effective uses of e-mail, automatization, and groupwork tools to schedule, monitor, verify, and expedite work.

After getting your car, you might be pleasantly surprised to see in the video clip that came with it how the manufacturer used the Information Marketplace long before you came into the picture. Designers, marketers, and managers worked in teams to design the car with proprietary groupwork tools and software modules provided by their company's information infrastructures. During the design phase, besides seeking one another's advice and experience, they collected information about the specifications of the most popular cars to date and the shifts in demographic interests of likely buyers. They carried out many what-if simulations, testing technical and marketing queries and proposals before deciding what they would accept and adapt. The vehicle destined to become your car, like other manufactured goods, was extensively prototyped and mock-tested against different terrains and markets. Managers earned their promotions by asking questions of these prototyping and simulation efforts that, in retrospect, turned out to be the *right* questions.

Much more was involved, but the company did not brag about it in the video. By accessing online user groups and consumer reporting services, you found out yourself that the company spent an unnecessarily huge amount of money in the visual stage of design, providing their engineers with goggles and haptic gloves so that they could experience what they would have seen and felt inside or under the car whose design was still on the (virtual) drawing boards. You also discovered the kinds of service problems previous customers of this model had to deal with. Just the same, you determined that the foibles and weaknesses were no more pronounced than at other car companies, so you decided to go for it.

Your video clip did inform you with great pride, though, that the company will continue to observe your car through the Information Marketplace. Anticipating mounting pressure from competitors, your carmaker pioneered a guardian angel type of software that resides partly in the car's computers and partly in the manufacturer's service system. It is guaranteed to alert you about regular maintenance, advise you about avoiding recurring yet preventable problems, some caused by your driving habits, and record and monitor malfunctions, accidents, repairs, and other major events in your car's life . . . all the way to its resale or final trip to the junkyard. Your manufacturer claims that this life-cycle information will help you maintain a healthy car. The company didn't say so, but you know that it will also be used to trigger the marketing of accessories and services to you, along with feedback to the company's engineers on how to improve future designs.

Our manufacturing example signals a broader trend. The increased premium on customized manufacturing, on having better and less expensive products ready faster than competitors, and on the power of the Information Marketplace to spread customer requirements into subassembly orders across the globe will force companies to move the final assembly of items closer to the sales office and the customer, especially if the items are expensive. This means a shift in the locations of a company's facilities. This trend will exert additional pressures, beyond those created by telework and groupwork, in moving labor away from urban areas to the countryside. In a strange way, the Information Marketplace may thus pull us toward the preindustrial practices of living in rural settings and purchasing customized goods and services, while retaining the economic benefits of the Industrial Era—which abolished these practices in the first place.

Services, Services, Services

Commerce, along with the entertainment and health care fields, will rapidly exploit information technologies. The evolution of the Information Marketplace will start from these three broad sectors, in part because consumer demand is great, the infrastructure capabilities are well matched to the demand, and there is already serious activity in these areas. Having discussed these three and manufacturing, we have touched on half of the industrial economy. What's left are a variety of financial, legal, and social services, which we tackle next, and of course government, which we'll address in the next chapter.

Finance and banking will be among the earliest services to join the Information Marketplace in a big way. They are already heavily computerized. Home banking and stock trading have appeared throughout the United States, enabling us to write checks, transfer money, and buy and sell stocks from our homes. The potential market is very large; Americans write some 70 billion checks a year and carry out a comparable number of credit card transactions. Most of these will become fully electronic in the Information Marketplace.

Other services will grow too. Standalone computerized real estate listings have been around for nearly ten years, though most have not fared well, because their providers have not kept the databases up-to-date with the rapid pace of transactions in the real estate market. Furthermore, by not having photographs attached to every piece of real estate they offer, and by not having all the houses and apartments for sale cross-listed, they have limited the utility of their offerings, even though they can search by cost, size, special features, and other factors. There is now a resurgence of these early and isolated efforts; many Web sites are springing up that show local and even national listings of homes, complete with photos and comparative data on schools, living expenses, and other important factors. It's hard to believe that people will opt for the old-fashioned and tedious way of searching for a place to live before first narrowing their choices by browsing a rich database that can instantly yield photos, videos, and descriptions, as well as a quick appointment with the real estate agent for a real visit. The Realtors who earnestly embrace the Information Marketplace will have quite an advantage over those who don't.

The Realtor's key task is to match a buyer's shelter needs to available houses. This is but one of a multitude of matchmaking services that will emerge in the Information Marketplace. Seeking and offering employment is already a major activity on the Internet. Some companies report better recruitment experiences through the Internet than through traditional approaches. Buying and selling lawn care, home repair, graphics services, editing services, financial management, legal advice, companionship, and much more will surely grow, in local and global settings. You will not want to look over all the world's ads for used lawn mowers, but you would undoubtedly maintain such a broad outlook if you were shopping for a large boat or a financial advisor.

Insurance companies will increasingly undertake within their own organizations this process of matching customers' needs to available insurance instruments. Major insurance companies have already reorganized their old divisions such as life insurance, medical, and automobile into a more integrated ensemble of dovetailing services that strive to match the needs of an individual, family, or company as they go through their normal stages of growth. Insurance companies will try to step up this integration themselves as they reduce the number of field agents who knock on doors in favor of direct marketing and service from regional centers. Other solutions may emerge, however, as agents recast themselves as insurance customizers, tailoring customers' coverage to their particular needs by drawing from the various offerings of several large companies.

Investment and brokerage companies will also use the Information Marketplace extensively to match needs with available financial instruments. With the increasing automatization of all direct electronic transactions involved in buying and selling securities, new instruments, new trading methods, and other significant changes will emerge. So much money flowing so rapidly among so many people will create a more efficient world market, on the positive side, and peculiar new instabilities and gamesmanship, on the negative side—perhaps bigger and more painful than the stock market crash of 1987, which was instigated largely by poorly administered program trading. This is an area where we'll have to be vigilant so that we don't create our own catastrophes through the very dynamics we will introduce in our quest for greater efficiency.

Legal services will assume a new face, too. Lawyers and laypeople will have increased access to information about related legal cases and new information as it emerges in depositions, no doubt in hyper-organized form. Law firms have already begun to farm out legal research in small units that can be handled by part-time lawyers and paralegals in lower-cost localities. Shopping for lawyers in countless specialties will be driven by better access to information about how individual lawyers fare in litigation. Legal advice will proliferate online, too. For example, in 1995 Steven Fuchs, a divorce attorney in Newton, Massachusetts, posted a home page for his firm on the Web. It brought more clients, but it also triggered inquiries from other divorce attorneys who wanted to post pages too. Within six months Fuchs and five other attorneys across the country who cover the gamut of divorce law had erected DivorceNet, which offers advice on virtually all the legal aspects of divorce cases, from paternity testing to mediation.

Noble Quests

Many of the Information Marketplace forces that will affect manufacturing companies and service providers will affect social and nonprofit organizations just as significantly. After all, the information work these groups must do is largely the same. We could discuss the specific changes for all kinds of organizations all day, so let's just pick two categories that show a range of possibilities. The two involve noble quests, one for spiritual wholeness and the other for scientific truth.

Religious organizations will be affected by the Information Marketplace in at least three ways. First, every church engages in a great deal of office work like organizing events, scheduling services, maintaining membership lists, soliciting funds, printing bulletins, and sending circulars to branches. The Information Marketplace will ease these tasks in the same way it will help commercial office work. Second, religious groups do a good deal of proselytizing. The Information Marketplace is ideally suited for spreading the word; the ability of each church to reach hundreds of millions of people with information about their beliefs and functions will widen the possibilities for affiliation. Of course, being visited daily by different virtual missionaries may be no different than being accosted by hordes of virtual telemarketers. People should have the right to shut out these intrusions as well.

Telework and groupwork will enable religious organizations to bring spiritual help closer to those who live in rural settings or who may be too infirm to go to church. These people should be able to listen to and actively participate in tomorrow's church services. In effect, because almost all churches build on the idea of a community, the Information Marketplace will help in the same way that it extends real communities to virtual communities. As in the case of education, however, the improvement will be felt mostly as an extension of the physical community, with totally virtual religious membership and fully virtual churches being rare and confined to cases where there is no other alternative.

Having said all that, it is important to note that we cannot foresee any ways in which the Information Marketplace will affect the spirituality of people one way or another, except perhaps indirectly by exposing many more people to the various options for spiritual fulfillment. Shortly after the mass suicide of a California cult group that thought it would board a spaceship hiding behind a comet and that used a Web site to promote its beliefs, I was asked by the news media if that's what we should expect from the Internet. I told them that these people probably had orange juice in the morning. So why didn't they ask me if that's what we should expect from orange juice?

Swinging from pure faith to hard logic, we move to the science and technology enterprises, which range from universities to industrial R&D divisions. The scientific community was one of the first to embrace networked computers: recall that NSFnet—the National Science Foundation's network—was a crucial step in the evolution of the Internet, and that Tim Berners-Lee invented the World Wide Web to enable physicists to read one another's papers wherever they happened to be. Though much of the rest of the world has just begun to exploit the Information Marketplace, the scientific community has used it routinely for years.

As information infrastructures improve, scientists and engineers will go beyond e-mailing their writings to one another. They will exchange more images and drawings, audio and video snippets that will expand the range and increase the quality of their interactions, for example, through improved data visualization. Some of these new modes may let distant colleagues watch or even instantly reconstruct an experiment by linking together remotely located but identical laboratory equipment. Environmental researchers, besides sharing ecological data sets about the environment, will also be able to access remote environmental sensors anywhere in the world. Already, NASA's satellite image database has been a gold mine for these researchers. Knowledge repositories should also become more widespread and, we hope, better organized.

Simulation will become more widely used, and powerful, specialized computer systems will become as coveted as today's particle accelerators. For example, by harnessing together a few thousand of the fastest computers, we should be able to construct in the next decade a virtual wind tunnel in which we can extensively, accurately test new aircraft designs before any actual craft is built. The same will happen with ship, spacecraft, and automobile design. Already there are simulators good enough for a portion of these tasks, but full-confidence simulation is difficult and very expensive to achieve due to the huge computational requirements. As computers improve, simulators will get better and the Information Marketplace will let scientists and engineers throughout the world try them out and share their results.

In their most advanced form these super-simulators can act as a new breed of *virtual telescopes and microscopes* that extend our probing to regions of space and time that were heretofore inaccessible. Consider, for example, the collision of galaxies—not an easy experiment to rig unless you are omniscient. But with the right kind of supercomputer capability you can simulate the crash, because the equations of motion for celestial bodies are well understood.

This has already happened. It was exactly with such a virtual telescope, the Digital Orrery, that Professors Gerry Sussman and Jack Wisdom of MIT discovered chaotic behavior in the motion of Pluto. They then confirmed this same behavior in the motion of all of the planets using a second MIT-built virtual telescope, the Supercomputer Toolkit. These "computational observations" resolved a centuries-old problem about the long-term stability of the solar system. Because of the importance of these results, the Digital Orrery has been placed in the Historical Scientific Instrument collection of the Smithsonian Museum of American History in Washington, D.C.

In mathematics, inexhaustible supercomputers have been used to search for and test new hypotheses, probe for new results like the largest known prime number, and check the veracity of complex and tedious proofs. In biology, they are being used to map the human genome and to simulate the three-dimensional shapes of genes toward helping us determine their biological function. Computers are also used as "virtual microscopes"—in drug design by simulating the life cycle of a virus to test how various inhibitors might affect it, and in material science to search for new composite materials with a desirable set of physical properties. In physics, they are being used to test new theories against experimental results, where the equations are generally simple but the calculations are immensely tedious and complex. As with the use of physical telescopes and microscopes, physicists report that after solving these giant problems they gain a better understanding of how things really work!

This list barely touches what has been done or is being done. There is no telling what novel insights and discoveries will arise in these and other disciplines through powerful virtual scientific instruments based on computers and through extensive use of the Information Marketplace for scientific purposes.

Pan-Organizational Changes

To this point we've considered all kinds of specific changes that will take place in different types of organizations as a result of the Information Marketplace. Let's now pull back and examine some of the most significant changes common to all organizations.

As we saw in the car purchase scenario, it is quite certain that the new techniques of visualization, product and market simulation, rapid assembly, and lifetime product monitoring will become increasingly important tools of tomorrow's business organizations. It is also a safe bet that e-mail, groupwork, and other middleware tools will rise in use among all organizations, improving communication and logistical coordination. These same tools should also result in increased reliability and speed, and lower cost, in whatever functions an organization performs.

The Information Marketplace will also increase the prospects for forming *virtual alliances* across hierarchical lines within an organization, with its suppliers and customers, and between sister organizations. We might reflexively assume that organizations will therefore become flatter and that all this distributed intelligence will conquer even the most difficult organizational problems because of the informality of the medium and the richness with which it can connect human resources.

Yet we should temper our optimism with the scary observation that organizations have had telephones for a century and that anyone could have picked up the phone and formed these alliances long ago. Some organizations may always be stifled by the isolation their corporate culture breeds. Faced with problems in manufacturing in the 1980s, managers at U.S. automobile plants sent memos up and down the chain of command, resulting in huge manufacturing delays of two-to-one compared with Japanese manufacturers, who met in teams across hierarchical lines. What will make an organization's employees who have not elected to team up with others in person or over the phone don virtual reality goggles and suddenly become cooperative team workers of the future? Will it be the novelty of it all? Doubtful. Sooner or later the novelty of a new technology fades and what survives within an organization is based on more fundamental factors of utility.

Human emotions and foibles have a huge effect on all professional exchanges within an organization. Solid bonds or rifts among employees, the boss's mood, motivation or lack thereof to achieve goals, passion, greed, jealousy, and altruism are all at play in any human organization. The Information Marketplace will have greater impact on organizations if it can effectively handle these subtle links among humans along with the more straightforward exchanges of information. Can the Information Marketplace pass—that is, transmit and accommodate—these intangible human factors that so profoundly influence our decisions and actions? And can it steer them to the good? Not quite and not unless people want it to. And even if the people are willing to use their new machinery toward these good purposes, there are still some human forces that simply cannot pass through the Information Marketplace, as we will see at the end of the book. Thus, the Information Marketplace must be supported with all the traditional methods for building human bonds, including face-to-face, real-life experiences, if it is to serve organizations as more than a high-tech postal system.

The Information Marketplace will undoubtedly alter the familiar organizational terrain in other ways that are harder to predict. Perhaps *remote work* from our homes will become so prevalent that the whole balance between cities and suburbs will be disturbed. As Bill Mitchell, dean of architecture at MIT, points out in his book *City of Bits,* such a transition would call for a shift of the physical workplace from the office to the home; an increase in restaurants and other personal services in the suburbs; and a corresponding decrease of the same services in the city. As we will also soon discuss, this demographic shift may also create a new breed of citizens, split between their city and village identities.

It's often said that the companies of the Information Age will have fewer people, that workers' voices will be heard more easily in the executive suite, and that teams of people will be easier to pull together and disband. Indeed, it is even said, with some breathlessness, that people will be able to assemble a company overnight across the globe, carry out the purpose for which it was assembled, then disassemble just as quickly.

It is certainly possible that organizations will have fewer people, especially because automatization is likely to increase human productivity. The notion of *instant corporations* is harder to swallow, because a great deal of a group's power rests in its people and especially in the relationships among them. Relationships cannot develop instantly even if the mechanics of getting people together can. If there is any chance for instant organizations, they will have to be formed among people who already know one another; we all know how old friends who have built trust over many years can reach momentous decisions in brief telephone conversations. Instant task forces within existing organizations, as in our example with British Petroleum, will happen. Instant organizations of people who have never met their peers, let alone built some mutual trust, won't work! Early results from British Petroleum confirm this observation.

Another probable organizational development is the evolution of "expert centers" staffed by groups of related experts capable of high-quality, high-speed work at very competitive prices. Instead of only using their voices, as is the case with today's call centers, these people will have access to all the resources and tools of the Information Marketplace and will be able to deliver information and information work to anyone anywhere in the Information Marketplace. In some cases the experts will reside in one physical location, but the people that make up most centers will be distributed throughout the world. Yet they will present a single, local organizational face to anyone who contacts them. We can imagine lawyers from different countries forming a center that specializes in international law. Personal and financial advice centers, medical diagnosis centers like the one with the Sri Lankan doctor, how-to centers, purchase collaboratives, and many other expert groups will emerge too, creating new alliances across the world. The fundamental economic force behind these truly new organizational entities will be the delivery of greater expertise at lower cost than was possible in the pre–Information Marketplace era.

We will also witness the rise of "work centers," whose demographic effects will run counter to full decentralization. Though work may move away from the large corporate centers to local communities, this does not automatically mean it will move into our homes. Work centers would be physical facilities in our bedroom communities where we gather to do our jobs. These co-ops or intermediary firms would have high-quality, high-bandwidth links to the Information Marketplace; sophisticated groupwork software; all kinds of human-machine interfaces; hyper-secretaries; cafeterias and other office-related amenities; and even day care facilities. Workers from different companies, and those who are self-employed, would go there to work. In addition to offering a wealth of office services, the centers would provide separation from the home and its distractions and a social environment with other workers, which people seem to like and even need. With these new entities in place, employment may ultimately become a mix of old-fashioned work at the company, work at a local work center, and work at home. Once again we see the Information Marketplace pulling against the urbanization that was caused by the Industrial Revolution as people swing back toward the rural communities in search of healthier, less expensive places to live away from pollution, crime, and the other ills of the world's large cities.

The Information Marketplace will have a very big impact on workers who, for whatever reason, can't get to where the jobs are. Its space- and time-bridging effects will allow disabled and home-bound people, parents of young children, and others to produce results that are indistinguishable from those produced by employees who can actually come into the office full-time. The Information Marketplace will do much to erase any practical basis for the reluctance many employers now feel toward hiring people who don't fit their standard employee profiles. The new legitimacy of home-bound and part-time work will also enlarge the pool of potential employees and expand the job market for people whose employment opportunities will no longer be restricted to the area in which they live. Thus the Information Marketplace will level differences in employment and unemployment across regions and nations with serious repercussions on the distribution of labor and wealth worldwide.

As for handicapped people, the potential benefits do not stop there. Already in Europe there is a large database called Handynet that documents the various aids available for people with different disabilities. Other matching and referral services are sure to arise. Moreover, bringing handicapped people into the ranks of productive workers benefits not only the economy but our society at large as these people fulfill their human desire to feel needed, useful, and equal to their peers. The Information Marketplace may be the healing force that removes the "handicapped" distinction from the organizational vocabulary.

Knowledge Fever

For all the organizational benefits it offers, the Information Marketplace will also set up a few traps we should avoid. One of the most glaring examples became apparent to me one day when I was attending a conference on information and work. I encountered a woman I didn't know who had made some clever off-the-cuff remarks during the question-and-answer period. So I approached her and introduced myself. She said, "Hello. I am the chief knowledge officer for XYZ corporation," and paused for my reaction. "What do you do as a knowledge officer?" I asked, with genuine interest. Her cheery disposition changed subtly to friendly contempt, as her thoughts seemed to flash before me: *If this guy doesn't know what a knowledge officer is, do I really want to meet him?* She responded with an upward toss of her chin: "I manage the company's knowledge assets." The devil in me rose, and I retorted instantly, "Do you also manage the company's knowledge liabilities and knowledge balance sheets?"

Her false smile now vanished, and she asked if I was serious or jesting. I explained to her that I really had no idea what a knowledge officer did even though I was knee-deep in the Information Marketplace. I further volunteered that I was leery of coy metaphors, for they tend to obscure real meaning with all the baggage that comes with the old formats they echo—in this case, financial asset management. She retorted with tactical acumen that she had no idea what the Information Marketplace was and asked me to explain.

The increasing use of these kinds of titles and other terms like *knowledge capital* signify the importance that corporate America is beginning to place on the role of information. Since we have a chief executive officer, a chief financial officer, a chief operating officer, and even a chief information officer, why not a chief knowledge officer too?! Aside from the somewhat pedantic notion and centralizing mind-set that every function needs a chief somebody, the first real question is whether there is a difference between information and knowledge. A common misconception holds that there is. Not so. Any information can become knowledge if a user finds it useful or interesting. So, for a business, knowledge is any use of information that helps the organization. But for most futurologists, their preoccupation with knowledge carries a belief that something loftier and much more valuable than the information itself is at play that, if captured, will place an organization on top of its competitors! What hype.

Since time immemorial, knowledge has been a key component of all human activity, and certainly all business activity. And as every student of "best practices" knows, the world's top corporations are characterized by knowledgeable employees who are aware of the broader corporate goals and issues and who know intimately the immediate business matters that surround them, whether they are in sales, marketing, engineering, administration, or research. I cannot help but recollect the "knowledge assets" of a Greek friend, a shipping tycoon today who wasn't always so wealthy. He had bought two of the shallowest draft ships in the world, all he could afford at the time. When he got a request for a cargo shipment to or from a shallow harbor, he would first check to see where his competitors' shallow draft boats were. If they were available, he would bid low. If not and the client was in a hurry, he would bid three times as high, knowing that he was the only game in town. The clients were incredulous at these fluctuations, but they had no alternative. Conquered by knowledge, they paid up.

"Knowledge" is not a few selected jewels kept in a safe and managed by a specific individual. It is a myriad of tricks, routine procedures, facts, contacts, and other nuggets that are possessed by every single employee from clerk to CEO, at every location and time the company does business. It's ubiquitous, like air and water, or perhaps like reading and writing and remembering. Every employee needs these things to function. So where is the company's chief air, water, reading, and writing officer?

The point, in case it got lost in my tantrum, is that if a company is to compete effectively in the new world of information, all of its employees must feel comfortable exploiting whatever aspect of the Information Marketplace helps them carry out their work better—from buying paper clips to setting the stage for merger negotiations. We have had CEOs, CFOs, COOs, and CIOs because these activities have traditionally been centralized. By contrast, the use of the Information Marketplace is most effective when it is fully distributed. Now if a knowledge officer is the person who will help educate a company's employees or who will focus on what employees do so they are able to better exploit the Information Marketplace, then I wholeheartedly approve of the position.

Whether or not chief knowledge officers become common as the Information Marketplace blossoms, it's a sure bet that chief information officers and their information technology (I.T.) teams will continue to be around. At one time, these specialists were the only ones dealing with information and its processing. With the changes ahead, in which almost every business employee will be dealing with information, the question arises, What might these I.T. people do? They should manage the company's shared information resources—the common data and procedures that everyone needs and the organization's shared infrastructure tools. For example, they should help establish e-forms within their own organizations, which should be easier than establishing e-forms between organizations. Tomorrow's I.T. team will handle a smaller fraction of the organization's total information activities than its ancestor—today's I.T. team—because so many more people within the organization will be using the Information Marketplace directly to carry out their work.

Finally, a word on outsourcing I.T. and other organizational functions is in order, because outsourcing is becoming very fashionable in business. The emerging argument goes something like this: Each company should try to be best in the world at what it knows how to do best. It should then outsource to other expert organizations the rest of its activities because it can never compete with the experts in these endeavors. This apparently sane argument is faulty. It typically omits consideration of the *interrelationships* among an organization's various activities, which are as important as and in some instances more important than the activities themselves. Never mind whether this is a valid business strategy or a fad. The Information Marketplace can support outsourcing of almost any part of the organization, because such outsourcing will almost always involve services and therefore a great deal of information.

There is one organizational function that should not be outsourced, however. It is I.T. itself. Information will be so intertwined with employees' activities and organizational startegy that outsourcing I.T. would be almost like outsourcing all the firm's employees. On the contrary, because the way in which an organization uses the Information Marketplace will be a powerful determinant of that firm's overall competitiveness and success in the world arena, it had better keep all this critical work in-house.

Accountability, Egalitarianism, Responsibility

The Information Marketplace will create structural changes across organizations. It will also affect human behavior within organizations.

Employees, indeed any member of any organization, will find that the Information Marketplace is likely to increase their accountability. The infrastructure makes it easy for bosses, shareholders, customers, and group members to monitor what individuals promised to do, or are expected to do, based on standing procedures. It will be difficult for someone to tell you that your check is in the mail when it takes only a few seconds to credit your account with money due you. It will also be difficult for someone to give an excuse like "I didn't know about it" or "Nobody told me" or "I tried but couldn't reach you" when written, verbal, and even visual instructions have been digitally signed and left in e-mail repositories. We saw earlier how schools can post homework so parents can see through their children's excuses. It will be no different with adults in business, and it will improve a firm's efficiency, even though its employees may be unhappy to have their individual work so open to inspection and critique.

Some idealists believe that the Information Marketplace will increase egalitarianism, equalizing the differences among people in organizations, both along and across ranks. This does not seem as plausible to me. More communication does not automatically ensure greater equality, at least in the short term. The telephone dramatically improved communication long ago, yet it did not seem to affect egalitarianism. A dictatorship is just as likely as a democracy to become more powerful by using the Information Marketplace to ensure that the dictator's orders and wishes are packaged with the right spin and communicated rapidly and effectively to the people. Surely, increased communication would allow the people to conspire and fashion revolts as well. But the telephone didn't help topple many dictatorships, typically for fear that conversations would be monitored. The same could happen in the Information Marketplace.

Wouldn't cryptography ensure that no one could intercept communications? It might, but even if it did it wouldn't necessarily disguise who was communicating. As we will see, over the long run the increased communications of the Information Marketplace will favor freedom more than oppression because of the power of shared information. But it does not quite follow that it will level the differences between the employees of an organization.

The most important behavioral change that the Information Marketplace will bring to bear upon an organization's employees will be a shift of responsibility further down the corporate ladder. Individuals will be able to solve larger problems with new tools before going to their bosses. They also will be able to handle problems and opportunities as they arise more quickly. Organizations will surely move in this direction, because it raises their effectiveness, lowers their costs, and speeds up their reactions to clients' needs and to competitive threats.

This raises an important point, however. For an organization to extract this increased decision power from its people, it will have to provide them with more knowledge about why some things are done and who does them, and why certain decisions are made and who makes them. The smart organization will purposely use the Information Marketplace's middleware tools to give its people the power of this knowledge. Automatization procedures will capture the substance of critical standing procedures. Groupwork will summon expertise to the right place at the time it is needed. E-mail will keep the organization's members apprised of the latest factors and debates affecting the organization. And organizing and finding aids, along with databases, will keep people current with all the valuable information they need to have.

Increased responsibility and knowledge throughout the organization will call for more on-the-job education, for it is the nature of organizational knowledge to change with time. Here again the Information Marketplace can help with training aids, simulators, and the pairing of teachers with students, wherever they may be.

Industrial Performance

With all the potential changes that we have discussed in this chapter and the many others that will undoubtedly come about when the Information Marketplace is used by the world's organizations, we are tempted to ask some simple questions: Will all this make a material difference to industrial performance? Will it give us the ability to manufacture better, less expensive, more reliable, and faster-to-market products with fewer workers?

Many people think so, especially the prophets of reengineering and downsizing. To get beyond the level of sheer belief to an understanding of how the Information Marketplace might help, let's first identify the factors that we know are responsible for improving industrial performance. Then let us see how the Information Marketplace might affect these factors.

In the late 1980s, MIT sponsored an extensive study of industrial performance called Made in America, which I had the good fortune to chair and which resulted in a popular book by the same name. Nine teams studied nine manufacturing industries at the sectoral and individual company levels through interviews of the entire range of employees, from the shop floor to the executive suite. The goal was to identify industrial strengths and weaknesses within a sector. Each team was unaware of the other teams' work. The independently obtained results were then compared sector against sector to identify in the common patterns that emerged the common strengths and weaknesses of U.S. industrial performance.

Between 1991 and 1995, France, Sweden, and Japan carried out comparable studies. The culmination was an international conference at MIT where these conclusions were combined with those from the United States under the leadership of Professors Richard Lester and Suzanne Berger of the MIT Industrial Performance Center. The conference produced a grand conclusion on the most common strengths among the world's top manufacturing firms. We can think of these factors as the best internationally shared industrial practices:

> A carefully nurtured workforce, made up of broadly well-educated and continuously trained workers who are given greater responsibility and are properly appreciated and rewarded.
>
> Greater cooperation within companies, with suppliers, with competitors, and with government.
>
> A focus on manufacturing through mastery of new manufacturing technologies, emphasis on processes, and continuous, relentless improvement.
>
> Living in the world economy by knowing and caring about the interests and habits of others, by shopping for the best technologies and the best suppliers worldwide, and by comparing practices to the best competitors worldwide.

A further overarching conclusion was unanimously reached:

> A major break has taken place in the world from the traditions of Fordism, where mass production reigned supreme and individual workers were a cost factor to be minimized. The new approach to manufacturing emphasizes people, and teams of skilled and continuously educated workers who share responsibility, are networked, and are self-governed. Compared to the old ways of manufacturing, these workers produce smaller numbers of more customized and novel products that reach the market faster, at lower cost, and at a higher quality.

This new manufacturing ethic, with its emphasis on people, was so prevalent around the world that the various national teams independently invoked a special name for it. The Swedes talked about the development of *human capital*. The Japanese dubbed it *humanware*. The French called it *Toyotism*. And the Americans called it *new economic citizenship*.

Let's now examine what the Information Marketplace might do to these proven factors of superior industrial performance.

The Information Marketplace can help bring workers a continuous, timely education. New technologies can also enhance cooperation through groupwork. And they can certainly help workers focus on manufacturing, through joint design sessions with marketing and manufacturing personnel and other experts and through extensive simulation and testing (torturing) of products and services well before they are launched. Furthermore, the Information Marketplace gives each company and industry instant access to its global neighbors; it will be difficult for any company or industry to find an excuse for not comparing itself with the world's best as part of its quest for even better products and services.

So we may gratefully conclude that the major changes that we have discussed here can indeed lead to increased industrial performance. We must be careful, however, not to ascribe to the new world of information unrealistic powers. The Information Marketplace offers no automatic relief from industrial performance problems, nor the guarantee that best practices will be achieved. Most important, it addresses only indirectly the overarching new factor of top industrial performance—the profound emphasis on people, and teams who share responsibility, are networked, and are self-governed. So the organizations that nurture, respect, and appreciate their workers most will have perhaps the single greatest advantage over their competitors. These attitudes will become even more important as the new world of information unfolds, because the employees will own, through their greater knowledge, a larger share of the means of production.

10
Government

Internal Changes

The world's governments are an important category of human organizations, with special needs and special powers. They will be
transformed at two levels: within nations, and among them. We'll
tackle the intragovernmental issues first, and take on the intergovernmental exploration in the following section.

Governments need to communicate with their constituencies.
They thrive on the use of forms and structured information, and
they can always afford to become more efficient. In short, they're
excellent candidates to benefit from the Information Marketplace.
Already, many U.S. government RFPs (requests for proposal), a
good deal of canned boilerplate on what various agencies do, and
announcements of various sorts are available on the Internet. The
EPA's Toxic Release Inventory lists toxic substances released in
specified locations by specified companies and is available to the
public electronically.

In the other direction, several federal agencies accept proposals
from citizens and organizations via the Internet, especially the
Department of Defense. Already some 15 percent of American tax-
payers are filing IRS forms electronically. Software vendors have
begun to sell packages that link home banking and other financial
activities directly with the requirements for filing tax forms so that
a taxpayer can meet his or her obligations automatically with little
hassle. These kinds of information transfers from and to the government are destined to proliferate because of the economies and
conveniences they offer.

The world's elected officials will also use the new technologies
to collect commentary and opinions from their constituencies, to
assess their own chances of reelection, and limit the success of
their opponents. Many people are already accessing government

servers that provide and receive information on all sorts of issues. In the future we are likely to see the emergence of automated responders—advanced versions of those inhuman voice mail technologies I have decried throughout this book. They will emit soothing messages like, "Thank you for asking. Your questions are very important to us. We will ship you an answer as soon as we can. Please check your mailbox periodically." And they will infuriate people who realize that they are the cheapest, most expedient, least sincere way to handle their query.

Electronic polling and voting can easily be carried out in the Information Marketplace. Of course, people must be willing to participate, which they will be initially. However, they will soon see these vehicles as burdens if all kinds of local, regional, and national agencies feel they have a license to poll everybody on everything. On the government's side, the ability to instantly poll people's opinions also has potentially serious repercussions. In some cases, perhaps after a natural disaster, being able to carry out an instant poll of the affected parties is valuable in determining which resources they need most—food, shelter, or transportation. But in most other cases, instant and excessive polling may result in a government in which officials make decisions solely based on how they might play to the citizenry.

Polling can undermine the ability of officials to lay out, debate, analyze, and implement their plans. By turning to instant polls we would eliminate the captain and the officers and leave every shipboard decision to the passengers. That is tantamount to virtual mob rule. It's a bad idea, one leaders and the electorate should guard against.

Governments can also communicate with the people through large assemblies and town hall meetings. Very large meetings are impossible in the Information Marketplace, roughly for the same reason that they are impossible in person: you can't understand *anyone* when numerous people are screaming at you simultaneously (unless, of course, they are all cheering or booing you). Likewise, you can't possibly read millions of written positions in a delayed version of such a meeting. For numbers of people in the tens, hundreds, or even a few thousand, electronic bulletin boards are useful and will enable the posting of political positions and the conduct of structured debate, thereby becoming useful tools in tomorrow's electronically augmented democracies. President Clinton cleverly bridged this gap from the hundreds to the millions during his 1992 election campaign by holding "electronic town hall" meetings where millions of people could watch but only a manageable studio audience could attend as full participants.

Every government is a major buyer of goods and services and therefore a major candidate for electronic commerce. A powerful example is the U.S. Department of Defense (DOD), which launched in the mid-1980s an initiative called CALS (initially for Computer-Aided Acquisition and Logistics Support and more recently, with tongue in cheek, for Commerce At Light Speed) to standardize the technical information exchanges about manufacturing and supporting the many platforms, weapons, and other equipment used by the military. Since then CALS has been broadened to include business as well as technical data interchanges. The thousands of suppliers and other enterprises involved in transactions with the U.S. government are expected to use this system and other related systems like the Federal Acquisition Computer Network (FACNET), thereby forming a huge governmental information marketplace. This medium will be used to receive requests from the government; give proposals to the government; handle orders, invoices, and settlement procedures; maintain schedules and reviews; and do many other such activities. These processes have already begun reducing the cost of doing governmental business and should continue to do so, while also increasing the speed and effectiveness of the government's overall procurement.

Another way governments will use the Information Marketplace is to link their own agencies and offices. Throughout the world governments have already begun developing their own national networks, which typically link their ministries or secretariats and other allied agencies. The links are largely limited today to e-mail, memos, and some intergovernmental document access and transfer. The potential of these networks is far greater, especially as automatization begins to relieve government employees and the public from the intolerable burden of filling out forms and waiting to submit them and have them processed.

Law enforcement within local municipalities or entire nations will be aided by the Information Marketplace in many ways, whether it is identifying a suspect from a "wanted" list, tracking a perpetrator across national boundaries with good police group work, or simulating the patterns of repeated crimes to predict future transgressions. Piecing together information has always been a big part of police work. The Information Marketplace will make this venerable activity easier and better. The FBI has put its Most Wanted list on the Web and reportedly is getting better results. More reporting of crimes, better surveillance where warranted, and improved crime prevention through the distribution of educational information may all come about. Obviously, the juxtaposition of "information" and "law enforcement" conjures up images of George Orwell's novel *1984*, in which the State snoops on the citizenry. We'll get to this issue later, in our section on Big Brother.

War and Peace

Once governments really begin using the Information Marketplace to alter their internal practices, they will be a short step away from improving intergovernmental activity. The world's governments will begin to link their networks to coordinate trade, improve world health, set all sorts of standards, carry out diplomacy, cooperate in crime control, increase tourism, and much more. These activities will gradually and inevitably lead to an "international governmental Information Marketplace"—a new channel of intercommunication among the world's governments and their people that will augment the traditional channels of trade, diplomacy, tourism, and citizen exchanges. At first, the new medium will become an informal coffee klatch where across national boundaries people will discuss with their governments and with one another topics of mutual interest. Government employees and businesspeople will participate in these informal exchanges. This is promising even at the informal stage, because any new channel of communication among the people and organizations of this world is likely to contribute to increased understanding hence greater peace. Later, more formal processes may be introduced as certain informal exchanges are found to be common and useful.

Regardless of these prospects for computer-aided peace, governments will still spend a great deal of time and resources on preparing for war. Since the end of the Cold War, the U.S. military enterprise, like others, has turned its attention to the "clever use of information technology," as they call this process, to enhance training and readiness with less and less expensive military hardware. The tank simulator we discussed is a prime example. A guardian angel for vehicles is another currently being contemplated. Governments have also recently started worrying about the vulnerability of their nations to information warfare—malicious attacks on their critical military and civilian information systems, such as those controlling the power grid, the telephone network, the air traffic control system, and the social security system.

Instant networks are another way the Information Marketplace will help tomorrow's military. These are designed so that a complete network can go up in a few minutes to link military units in some remote trouble spot with the generals and specialists back home and with several standby and logistics support teams. This instant network can supply information to the soldier as intelligence accumulates and to the command post about what is happening in the field. Guardian angels could be used to help with medical problems that arise in the field—in fact, MIT's very own guardian angel research was originally funded for this application by ARPA. E-mail would handle messages at multiple levels of security. Groupwork would bring the right minds together to confront unexpected situations with new tactics and changed strategies. It is even possible to let soldiers "see" behind hills or detect bunkers buried in the sand by using augmented reality to superimpose upon the real scene visual information from supersensory interfaces (radar, infrared, sounds, and so on) and from reconnaissance information.

If we can't stop fighting with one another, maybe we can delegate the job to robot soldiers fighting robot enemies, thereby destroying hardware instead of killing people! The idea is not so far-fetched: I once got a call from a friend in the Defense Department who wanted to know if our lab could build a pogo stick that could go bouncing around the countryside oblivious to hills and holes until it encountered an enemy tank, at which point it would simply hop atop the tank and explode. I explained that technically the problem could be handled, but we were not in the weapons business and he would be better off with a military equipment supplier.

The greatest military use of the Information Marketplace is likely to be in intelligence, however. After all, intelligence is nothing but the collection and analysis of information from informants, electronic eavesdroppers, publications, satellite images, rumor, and much more. An intelligence analyst, sitting comfortably in front of his sophisticated gear somewhere at CIA headquarters in Langley, Virginia, would be receiving all this information from all these sources as long as it dealt with his little area of expertise, say eastern Mongolia. If strange data began to appear, he would apply his own experience along with various what-if automatization procedures to test for familiar patterns and ultimately to assess what might really be happening in the Asian steppes.

Intelligence technology could take the most unlikely of forms. In the early 1970s the then chief scientist of the CIA visited our laboratory to find out what new technologies might be appropriate for intelligence applications. Each time we showed him and his team something new, he'd respond with an impatient, "Yes, yes, yes," suggesting he already knew all about it. At the end he asked, "Don't you have anything really exciting?" My predecessor, Ed Fredkin, winked at us and said, "Well, we could always build a small robot that would fit into the Moscow sewer systems. You could drop it in a manhole outside the Kremlin, and it would navigate its way, much like a cruise missile, through the sewage pipes, eventually reaching the General Secretary's toilet. At that point it could hoist up a periscope TV camera or explode or whatever!" We were all preparing to explode in laughter when the chief scientist looked at us somberly and said, "How much do you need, and how soon can you have a prototype ready?"

In principle, industrial intelligence is no different than military intelligence. The Information Marketplace is therefore ideally poised for civilian snooping applications. But the technologies that make intelligence good also make counterintelligence good. Military and civilian adversaries alike will engage in a series of measures, countermeasures, and counter-countermeasures to gain advantage through knowledge. Ultimately, these activities, despite their ostensible secrecy, will also give rise to greater openness as armies and companies get to know more about one another's goals and pursuits. In fact, information about organizations and their activities will become so pervasive that the boundary between surreptitious intelligence and genuine legal market research will become quite fuzzy. Hopefully the outcome of all this will be greater peace and improved products and services for all of us.

Justice and fairness could even be influenced if the world's judges have better access to related cases, equalizing disparities across nations. New case law on, say, the validity of a DNA-based defense could be instantly available to the world's judges and legal systems to consider and possibly adopt. Redesigners of the legal system in a country undergoing major changes, like China, could test rulings of their judges against those made in courts in other countries. With improved information flow across national borders, extradition would probably become easier. Unfair practices such as child labor would also be more readily identified. These potential changes give us reason to expect better-administered justice and perhaps even increased fairness across the world.

Privacy Fears

As we have discussed in the last two sections, governments will view the Information Marketplace as a means for improving their legislative and executive functions, including the pursuit of war and the preservation of peace—and they will undergo substantial changes to capitalize on these possibilities. But governments will also see the Information Marketplace as a complex and potentially dangerous force that will need to be regulated, much as air travel, telecommunications, and the stock market are regulated today. As a result, the reactions of the world's governments to the new technology of information may turn out to be as important an agent of change as the technology itself in the evolution of the world's global and national information marketplaces.

The Information Marketplace concerns governments on two fronts. One is its pervasive reach, which tends to ignore national boundaries. The other is the privacy that new encryption systems can give criminals and anyone deemed an "enemy of the state." Let's call them the *government fears of pervasiveness and privacy*, respectively. We'll tackle the privacy fears here; we'll assess pervasiveness in chapter 13 when we discuss one of the major social forces of the Information Marketplace—electronic proximity.

Government security agencies are of two minds. On one hand, they want *strong cryptography* (extremely hard to break codes) for their own military and diplomatic communications; on the other hand, they want *weak cryptography* to be used by potential adversaries so that the codes can be easily broken. When it comes to their civilian populations, governments are equally schizophrenic. They (or at least most of them) would like to ensure the privacy of communications among the citizenry, but they would also like to be able to break this privacy to monitor the communications of criminals.

When public key cryptography appeared, it changed the business of cryptography from an elite craft known by very few specialists to a do-it-yourself affair. The result was a shock felt by various nations' security apparatuses. Governments saw public cryptography as a threat to their political and military balance. The scientists who created these schemes, however, saw them as vehicles for furthering science and for ensuring computer security in a world where information was becoming increasingly dominant. The dramatic efforts of this tension are worth recalling because they illuminate some deeply ingrained governmental attitudes about secrecy, security, and the Information Marketplace.

Ron Rivest, as the principal inventor of the RSA approach, and I as his laboratory director, first experienced the tensions when I called our sponsors at ARPA sometime in the late 1970s to tell them about the exciting RSA public key discovery and to ask them whether they would fund research in this promising area. They became as excited as the rest of us and eagerly agreed to the funding. On our side, we were anxious to carry out this work because we were already envisioning a new world in which precious civilian data, especially medical and personal, would need to be safeguarded from unauthorized use. The night of that ARPA phone call I went to sleep happy with the sociotechnical prospects for this work and with the growing fame of our laboratory.

The next day I got a call from the same person at ARPA who had been so excited only a few hours ago. This time he was subdued. He said, "Sorry, Mike, we have no line item in our budget to support this research." I kept asking him to explain what he meant, and he kept repeating the same sentence, until finally it dawned on me that the National Security Agency (NSA), which is responsible for U.S. government cryptography, must have put pressure on ARPA to stay clear of this sensitive area.

During the following months, the NSA tried various ways to contain academic research in public cryptography. They offered generous research funds but only in return for our compliance with various restraints, for example, their right to review all our potential publications so that they might decide which ones could and couldn't be published.

To a university dedicated to the free flow of ideas, this was tantamount to intellectual decapitation. Acrimonious debates flared. At one point the NSA intimated that they knew all along about the RSA technology because of some old secret British research, and that the scheme was fatally flawed. Later they admitted that the flawed scheme was not RSA.

The NSA, trying in good conscience to do its job, finally concentrated on what we dubbed gray controls—policies mutually agreed upon by government and universities that were neither dictated nor prohibited by existing laws and regulations but that would limit what NSA saw as potential damage from proliferating publications. By then, several universities had accepted similar restraints, and we were coming under increasing pressure to comply. We felt strongly that we should publish our results, however, because there were few if any laws around to protect the rights of the civilians who would need this technology to secure their medical, financial, and other personal information. We also felt strongly that, notwithstanding its clout, its multibillion-dollar budget, and its acres of computers, the NSA had no business telling a university what to do with new ideas if national security wasn't at stake.

The NSA too felt very strongly about this issue—so much so that the agency's director became directly involved. I'll never forget the commanding figure of Admiral Bobby Inman in his sparkling blue uniform when he joined Ron Rivest and me for lunch in the agency's dining room to discuss this conflict. He is an impressively sharp man with a total recall for facts—a feat he achieves without notes. Though we respected one another, we did not agree on everything. He did not share our vision of an Information Marketplace where civilian data would have to be protected to the extent we advocated. We had a hard time understanding why the NSA was so disturbed by our research. Rivest and I finally came up with an idea that broke the deadlock. We at MIT would voluntarily send to the NSA drafts of our papers at the same time we sent them to our closest colleagues for comment and criticism, without asking the government's permission to publish our work. If they wanted to stop us, they could always talk to us and even try to classify our work as a potential danger to national security. The NSA agreed, and we along with several other universities have since adhered to this practice. I am happy to report that not once did we hear that further publication would be against the national interest in NSA's judgment.

Not every story will have such a happy ending. After all, MIT and the NSA were largely on the same side, worrying about different aspects of the same nation's security. Yet, the fundamental tensions of the MIT-NSA conflict will persist with regard to the Information Marketplace. On one side there will be a group of "desirable" people (the armed forces, the diplomatic corps, the civilian population) whose privacy the government will want to ensure. On the other side there will be a group of "undesirable" people (a foreign adversary or a bunch of criminals) whose privacy the government will want to break.

Today, no one disputes companies' rights to security in commerce; the world's law books are replete with commercial privacy and intellectual property statutes. Furthermore, the world's governments follow a variety of different practices for intercepting communications of citizens under suspicion of criminal or undesirable political activity. In the United States and several other countries, judges can issue wiretapping orders so that law enforcement agencies can monitor the phones of suspected drug peddlers and other criminals. What's so different in the Information Marketplace?

The most important difference is a huge increase in the volume of information that governments, competitors, criminals, or simply nosy people may be able to intercept. As we have seen, information could account for more than half of our business and many of our personal activities. The volume is so large and its potential penetration by unauthorized individuals so far-reaching that the overall effect amounts to a substantial qualitative difference that must be addressed by individuals and governments.

Fortunately, there are already several ways to address it. We can let strong cryptography grow on its own, as it is today, allowing people who wish to communicate securely to do so in the ways we have already discussed. This would mean that citizens would be ensured privacy, but governments would be unable to break into the Information Marketplace transactions of suspected criminals or foreign adversaries.

Or we might elect to use so-called *key-escrow* schemes. In one of these schemes developed by Professor Silvio Micali of the MIT Laboratory for Computer Science and his colleagues, everybody's private key, when created, would be broken into three numbers that would be sent to three trusted agents, perhaps three judges. If a judicial directive demanded it, these judges would authorize the executive agencies to use the three numbers together so as to reconstruct the key and break a suspect's code. The technology can guarantee that no single number or pair of numbers can be used to break the code. Simpler schemes involve two or even only one number placed in escrow.

In yet another scheme proposed by several researchers, every citizen would be obliged by law to save the key in his or her own records and surrender it to government authorities under proper court order, much the way tax and other documents are handled today. The government would then be able to decrypt suspicious past communications using that key. This scheme can apply to any cryptographic approach, and it requires the government to notify citizens before carrying out search procedures. In an escrow system, such notification is not essential and the interception might be done without the citizen's knowledge.

Other schemes will be legally or illegally employed throughout the world. This situation is not too different from today's physical security devices, which range from rice curtains to steel vaults. In this rich setting of security capabilities, there are no foolproof approaches. Government agencies will try to limit the radius of perceived damage, and civilian users will try to get the least-expensive system that gets the job done with acceptable security. Members of criminal syndicates will resort to their own forms of strong cryptography.

Of course, governments are not only interested in accessing criminal information, they also wish to ensure that various vital information systems we all depend upon, like the telephone and the power grid, are well secured. We should not let a debate about access slow down securing these vital systems.

All the security schemes we have discussed and their variations lead us to an important conclusion: *We have the technological means at our disposal to provide essentially any desirable balance between the conflicting extremes of offering privacy to one group while retaining the right to tap the information transactions of another group.*

This conclusion presumes that everybody agrees to utilize the selected approach. This, of course, is no more possible to enforce than any other of today's regulations, which are routinely ignored by criminals. We must therefore view this conclusion in the broader and necessarily imperfect context of law enforcement, which would include means for detecting and enforcing noncompliance.

The overarching question is ultimately nontechnical. The people and organizations of this world will have to decide whether they want to be the ones controlling the privacy of their communications, or whether they are willing to share this right with their governments in order to protect themselves from criminal and enemy activities. This trade-off should be debated by national legislatures and by ordinary citizens around the world.

Absent a decision to consciously and explicitly change our ways, we should ensure that today's balance between citizen privacy and governmental interception is preserved. After all, human nature shows no signs of changing soon! And technology itself is not forcing any particular human policy direction upon us. However, in doing so, we should be careful not to misread the consequences of any new policies we introduce to preserve the current balance. For example, requiring every U.S. citizen to use keys that are registered would shift the balance, because it would enable the government potentially to dip its hand into a huge category of business and personal transactions that are currently unreachable. Conversely, allowing strong cryptography to become the norm would shift the balance in the other direction, giving terrorists and all sorts of criminals a secure communications haven in the Information Marketplace.

We should be prepared, too, for the realities of not being able to reach consensus about the course we wish to follow. Together with an increasing premium on information and therefore on information security, these natural trends may well turn the world's tides toward strong cryptography and greater citizen-controlled privacy.

In the last six chapters we have touched on many details of our daily lives and our pursuit of pleasure, our more dutiful endeavors concerning our health and education, and our activities through human organizations, including government. We have seen that the Information Marketplace stands to affect many of these activities broadly and deeply. Throughout, we have also begun discerning some broader patterns. It's time to widen our view to the bigger picture. The third part of the book examines the deeper meaning and influence the Information Marketplace holds for our economy, our society, our behavior, and ultimately our history.

PART III

Reuniting Technology and Humanity

11
The Value of Information

End Game

Now that we have assessed the new computer technologies, and their effects upon us, we must try to understand the extent, the meaning, and the depth of the entire Information Revolution for all of humanity. The Information Marketplace will bring significant economic, social, political, and psychological change.

We will begin in this chapter by discussing the economic effects, which are driven by the value of information. This will lead us to some unexpected conclusions about changes in the gap between rich and poor, in employment, and in the ultimate reach of information.

We will then confront two major forces arising from the Information Marketplace that will augment human capabilities: "electronic bulldozers" and "electronic proximity." These new forces will relieve us of work and bring all the world's people immediately close to one another, which will be both a blessing and a curse. These pivotal capabilities and their consequences for human productivity, politics, culture, ethnicity, and humanity are the topics of the next two chapters.

Finally, we will explore what might happen to human relationships and the human psyche when the Information Marketplace meets head-to-head with ancient, unchanging human qualities that we all share. We will take in the whole human picture and, in doing so, discover that we are participating in a movement that may go well beyond the confines of information, to a "renaissance" of the way we view ourselves.

Let's begin with my favorite exercise—debunking a myth.

The Myth of Cheap Copies

Some people argue that compared to physical goods, the information that will flow over the Information Marketplace has essentially no value, because it can be easily replicated. They gladly explain their argument in vivid terms:

> If I have some potatoes and give them or sell them to you, then you have the potatoes and I do not. However, if I have tomorrow's weather forecast and give it or sell it to you, then we both have it.

They go on to argue that because information is easily replicated it proliferates and is not scarce. And because economic value is rooted in scarcity, information ultimately has little or no value. My favorite response to this argument is:

> If information is that cheap, why don't you copy the tax accountant next door and give me a hundred copies. Then I'll open a tax preparation office and make some good money.

My challengers look at me as if I have gone mad, and say, "Surely a tax accountant is not information!"

Proponents of the cheap copy myth regard information as passive, what we have come to call an *information noun*—a memo, a database, a picture, a movie. And they imagine that there are a number of people around (a market) interested in buying copies at some low price. Both views are limited.

To be sure, information in publishing and entertainment—books, musical recordings, videos—satisfy these conditions. To these you can add patents and other intellectual property. In these cases the argument of cheap copies is true. But as we have noted, these information goods represent a very small portion of the economy—perhaps 5 percent in the United States. A far greater amount of information comes in two forms. The first is all the customized information (noun) tailored to individual uses, like the files in your office and home and in all the other offices and homes around the world. The second is active information (verb), like the tax accountant's work on your tax return. This is the kind of information we have called information work, which we defined as the transformation of information by human brains or computer programs.

There are a multitude of human, machine, and human-machine processes that carry out useful information work: the design of a building; the negotiating of a contract; the advertising, billing, and settling that accompany every sale; administering medical advice; brokering—in short, the huge amount of office work that takes place in all businesses, services, and organizations. It is these human activities that groupwork, telework, automatization, and much of what we have been discussing will improve. With more than 50 percent of the industrialized workforce engaged in office work today, information-as-verb activities dominate the terrain of information.

The question then arises, What is the value of information work? Our assessment provides a simple and provocative answer: "There is no fundamental economic difference between the value of information work and the value of physical work!"

Information work is done either by humans with their brains or by computers with their programs, or by the two in combination. This is no different than physical work, which is done either by humans with their muscles or by machine with their moving parts, or by the two in combination. In both information and physical cases, if the work is produced by people, it requires the expenditure of some portion of their human lives, regardless of whether it's brains or muscles that are involved. And as we know, the work of both information workers and physical workers is valued through hourly or monthly wages. Teacher and plumber alike are compensated for the use of their lives. And they are compensated based on the same economic relationship: the availability of supply versus the world's demand.

If the information work or physical work is carried out by machines, then someone or some organization must provide the capital to buy the machines, be they computers or bulldozers. Thus, the generation of information work, like physical work, depends upon the traditional factors of production, namely, labor and capital.

But what about the software that does information work. Can't it be easily copied like the information nouns, corrupting our theory?

Only the shrink-wrapped variety at software stores and in catalogs can be easily copied. Although this is a sizable market, just like that for CDs and videos, it is a small part of the active information work that we'll see in the Information Marketplace, which will entail a great deal of customized software and intricately dovetailed combinations of human and machine procedures. This situation is not much different from the factory, where ordinary pieces of the process (assembly stations, motors, tools, conveyor belts) are intricately dovetailed with one another and with human procedures to produce unique products and processes that are hard to copy.

Imagine, for a moment, all the human, machine, and human-machine procedures of the world's insurers, banks, finance companies, law firms, manufacturers, and governments. Would you pay even one dollar to have any one of these procedures and software programs delivered to your house? Of course not. They are of no use to you. The question extends to customized information nouns, too. Imagine the billions of files in the millions of corporate and individually owned file cabinets and computers in the world. Do I have an offer for them?

Customized information, whether of the verb or the noun variety, is of no value whatsoever to anyone except its owners and a few closely allied parties. To be sure, the blueprints for a new car or the formulas for a new drug are information of great value to the organization that owns them and its immediate competitors. And they should be mightily protected from being copied. But they are of little or no value to the rest of the world, because hardly anyone will be interested in paying even a penny for the copies. And although there may be a resale market for such goods, it is likely to be small and highly specialized. The bulk of intermediate information out there consists of procedures and data that are far less exciting than these precious secrets. There is no market for these information goods! In a sense of grand poetic justice, whereas pieces of data are easier to copy than pieces of chocolate or cars, they appeal to far fewer people. Yet, customized office work and these valuable-to-a-few information goods and services are the bulk of what will be used, bought, sold, and freely exchanged in the Information Marketplace. So the peculiar property that some information can be easily copied at low marginal cost is essentially irrelevant to the huge amount of information (noun or verb, human or machine) that churns around in the world today, and that will therefore churn around increasingly in tomorrow's Information Marketplace.

The fact that most information is of little value to most people doesn't mean that specific pieces of information aren't as valuable as (or more so than) chocolate or cars. Popular computer operating systems and Web browsers are used by tens of millions of people and require huge amounts of creative and development work, maybe even more than the making of chocolate or cars. So let's appreciate them as much as we do physical goods, and let's not grant ourselves license to copy them just because it's easy, as some people and organizations do.

The pirating of software may abate as international agreements rise and prices fall. But there may be other ways to tackle the problem. In one radical scheme proposed by my colleague Steve Ward, a large number of software vendors agree to provide all their software for a rental fee, which they set periodically and which remains flat to each user of their consolidated service. You pay fifty dollars per month, and you can download from their Web sites *any and all* software that these companies make. In your computer's operating system, there is a secure "taxi meter" that records your usage of this software and reports it automatically to the service at the end of the month, when you credit them with your payment. The service takes in the total proceeds from its customers and pays the participating software vendors, based on an agreed-upon formula, according to the actual use of their software. Ward sees this approach as a conversion of software from property to a service that eliminates the incentive to cheat by copying. He also believes that the scheme will promote wide dissemination and communication of new software, as opposed to the current approach that discourages communication to protect the vendors. Ward thinks that the scheme can be extended to all intellectual property, which he sees suffering from the same pitfall. In its ideal extreme, this approach would require universal agreement that is nearly impossible to secure. It may well start, however, in software clubs and may grow bigger if it is proven successful.

A variant of the cheap copy myth must have penetrated our society, which continues to deny the equality of some information work to physical work. This is probably due to our agrarian heritage. Back in the days when hard physical labor was needed to toil the land, information work involved a small amount of reading and writing under the oil lamp, often done by the farmer's wife and children. Compared to the heavy work of handling the ax, the plow, and the farm animals, this was a relaxed activity that did not command the title of "work."

Perhaps this is why governments, clinics, and countless organizations, not to mention telemarketers, consider it their birthright to ask us to expend valuable portions of our lives filling out long, obtuse forms, walking through complex menus of choices on automated telephones, or patiently responding to queries, as if our lives are theirs to spend when they ask us to pull information together for their purposes! Surely, none of these people would have gotten away with the physical equivalent of such requests—boldly ringing your doorbell, handing you some wood, and asking you to expend a portion of your life to assemble a table for them, free! The day will come, not too far into the future, when we will revolt at these info-assaults and pass laws that forbid them, unless, of course, we give explicit permission to the contrary. The time is approaching for a "manifesto on human information work" that, like labor law, would govern what is and is not acceptable in the Information Age. End of tantrum.

There is one other growing misconception we should dispense with: that because information work derives from the brain, which is so far "above" our muscles, it is therefore a lofty endeavor that will somehow uplift the human spirit of the participants to a new order of goodness, culture, and understanding. No doubt this will happen in quite a few cases, and the Information Marketplace increases our prospects for doing so, worldwide. But lest we get carried away, we should sweep over in our mind's eye what the millions of office workers do every day throughout the world and ask if their work—that will permeate tomorrow's Information Marketplace—comes near to meeting these exalted goals!

The conclusions from all this discussion are clear. A small part of the new world of information—the writings, songs, videos, and software packages in popular demand—differ economically from physical goods in that they can be copied at low marginal cost. Most of the information out there, however, represents information work (verb), customized information (verb and noun), and human procedures carried out with these info-verbs and -nouns that is either difficult to copy or irrelevant to most people if copied. And, in economic terms, information work is largely the same as physical work, with both requiring labor and capital to be carried out.

On balance, the new world of information is economically closer to the old world of physical goods and services than is generally thought. Bits may be plentiful as sand, but, like sand, they are useless unless fashioned.

The Economic Value of Information

Now that we have established that information and information work have value, we will try to understand what that value is so that we may gauge the impact the Information Marketplace will have on the economy.

Information Theory, developed by Claude Shannon in the middle of the twentieth century, tells us that the information in a given situation is roughly the number of bits needed to describe all the likely outcomes. *The sex of our new baby will be male or female* is a situation that has one bit of information, because there are only two equally likely outcomes (1 or 0), which you can communicate with one bit. *The sun will rise tomorrow* has zero information, because no bits are needed to communicate a certainty.

This theory helps us determine how fast and how well we can communicate messages over noisy phone lines—for which the theory was created. But it becomes stretched beyond common sense when we apply it to what we do in the Information Marketplace. Great fame awaits the techie or economist who will fashion a new, useful theory for this universe. Until then, we will have to cope with more intuitive discussions to develop a plausible way for valuing information. So let's get on with an approach I have developed for that purpose.

Think of all the goods and services around us as being either *informational* or *physical*. On the information side, we have office files, newspapers, videos, purchase orders, the work carried out by office workers and computer programs, and so on. On the physical side, we have flour, bread, cars, restaurants, retail trade outlets, and more.

Now consider these goods and services and call them *final* if they are consumed by people or *intermediate* if they lead to further goods and services. On the physical side, bread is a final good because it is consumed by people, whereas flour is an intermediate physical good that must be further processed to lead to bread. Similarly, a restaurant waiter offers a final physical service, whereas a wholesaler carries out an intermediate service between manufacturers and retailers.

This division applies equally to information. A video like *Star Wars* is a final information good, whereas a mailing list is an intermediate information good that leads to something else, like the sale of a product. Similarly, a talk show host offers a final information service that is consumed by the audience, whereas a human movie editor offers an intermediate service that leads to a movie.

Because they are directly consumed by buyers, final information goods and services are economically similar to final physical goods and services; their value is based on the human desires they satisfy and on their scarcity—the familiar law of supply and demand. This is why a video copy of *Star Wars* is valued at a few dollars whereas a host like Jay Leno or David Letterman is valued at millions of dollars and may be too expensive to hire for a private function. As we've noted, the fraction of the U.S. economy that deals with final goods and services today is very small.

By contrast, intermediate information goods and services are a much larger part of the economy and are widespread, because they lead to millions of goods and services and because they include all office work. But because, as we have seen, intermediate information is mostly customized, its value is determined largely by the value of the goods and services to which it leads (the economists call the demand for these intermediate goods *derived demand*). We should also note that a company's intermediate information, with all the customized human office procedures and software that it entails, is not as immediately marketable as intermediate physical products, which tend to be more standard (like flour).

We now have all the background we need to answer the question, What is the economic value of information? *Information has economic value if it leads to the satisfaction of human desires. A small portion is final goods, which derive their value from supply and demand. By far the larger portion is intermediate goods that derive their value substantially from the value of the goods and services to which they lead.*

Before we go on to use this new insight, we should note that many activities that seem at first blush to be final information are actually intermediate. For example, even though we consume electronic shopping and legal advice, they may actually be intermediate to further actions, such as giving a gift or reducing the taxes we pay. Education may look like a final good, short term, but ultimately it is handy for earning a living, long term, so much of it must be viewed as intermediate. The value of a college education may be thought of as perhaps 3 percent of the additional total revenue it will lead a college graduate to earn throughout a lifetime of work. When these seemingly final information goods and services are called by their true name—intermediate—there is even less left in the category of true final information goods. Intermediate information goods and services are thus even more pervasive than they seem at first.

Some of the other ideal distinctions we have made also need qualification: real goods and services are not divided cleanly into physical and informational; each will usually have a mixture of these components. In fact, soon the world of business will start talking about the *information content* of the various products and services they sell. We can still value these products by valuing their components. Information may become temporarily overvalued, as human expectations rise around a crucial hot product or some coveted human desire. Once the hype subsides, however, people will again value the information according to the value of the goods to which it leads. It may also take time for certain kinds of information to realize their value. Several years may pass before the full revenue from a software program has been realized. That does not change the fundamental picture: whether a piece of intermediate information is viewed as a capital asset or as a revenue flow over the long run, its value is derived from the value of the goods to which it leads.

Let's test our way of valuing information on a couple of examples before we go on to use it as a basis for more momentous conclusions.

What's the value of a mailing list? A business that sells encyclopedias will value a mailing list of families in the Midwest that have children of "encyclopedia age" by discounting backward from the value of the additional encyclopedias that the list will help sell, reduced by associated costs and risks. So if the list is expected to bring in another $100,000 of revenue, $10,000 of which will be profit, it may be worth a fraction of that profit adjusted for risk and other factors—let's say 3 percent of that profit, or $300. That's how the company would figure whether the list is worth buying. Another mailing list, which catalogs midwestern families that could afford to buy such an encyclopedia, might have a similar value. Suppose now that an entrepreneur processes these two lists to develop a new list of likely candidates who meet both criteria. The company might pay perhaps $1,000 for this far better targeted list, because it will incur fewer costs to pursue fruitful leads. In this way, the information work performed by the program to mesh the two lists ends up being valued as well.

These lists would have essentially zero value to a restaurant owner. The information is only of value to the people whose desires it satisfies.

What is the value of a word processor? The value of a word processing program to you stems from the value of the documents it helps you generate. Their value, in turn, depends on the value of the goods and services to which they lead. If you are a poet who composes poems for your own pleasure, then you probably place upon the word processor some small, intangible value; you can write the poems just as effectively on a piece of paper, although the spell checker is handy and the poems look nicer in your favorite font. However, if you process claims in an insurance company, your pay and promotion rest on the speed and quality with which these documents are produced. The word processor's economic value to you and your company is great. The company might value this process sufficiently to pay a software house a few million dollars for a customized word processing system that rapidly and accurately produces claims documents, unfettered by the excess baggage of a general purpose word processor. Would you pay anything to write poems on that word processor? Surely not!

The value of information need not be confined to economic terms. What is the value of knowing that a dear friend is not terminally ill, as he had feared? Knowing your friend's diagnosis provides deep emotional value. And although you cannot assign a monetary value to it, the same rule holds: the value of the information is determined from the value of the intangible things to which it leads—in this case the love that you feel for your friend.

Armed with this way of valuing information, we'll now develop some conclusions about the new role of middlemen, the rich-poor gap, employment, and the potential economic reach of information.

Info-Junk and Intermediaries

With a billion or so interconnected computers around within a decade, and with each computer carrying a few thousand to a few million pieces of information, we will be surrounded by a mountain of data—somewhere between a trillion and a quadrillion files, programs, notes, lists, and other material. The great majority of it will be intermediate information goods and services, each one directed at a very limited number of consumers who will place value on it. For most of us, most of it will be a mountain of info-junk with zero value. All we want are the few nuggets that help us in the pursuit of our own desires and goals.

What would you do if you had to access all the catalogs and related information offered by all the music stores, magazines, and other suppliers in the world to order a single piece of music you desire at the lowest cost and fastest delivery? All this extraneous information would be irrelevant to you, getting in your way like a horde of shoppers blocking your path to the best music store with the right selection. It will certainly be annoying, probably even disruptive. And this is just one isolated person's isolated pursuit. Imagine the number of such frustrations you'd experience every day, and then multiply that by the number of other people out there in the same position.

To perceive the full magnitude of the info-junk problem, imagine the impact of all the physical junk that would be generated in a community of a billion people living close to one another. The Information Marketplace, by placing every person and computer a few mouse clicks away from every other person and computer, creates such a community in virtual terms. We would certainly not tolerate all the physical junk; likewise, in the Information Age, a great deal of effort must be expended to fend off and filter all the info-junk.

Automatization procedures in the form of "agents" will help somewhat, but we have already discussed the difficulties that will limit their broad impact. This leaves a lot of room for companies and individuals that will help us find and match services, like the financial planning service that helped our fictitious father select stocks, or the service that reported consumer experiences with JoAnne's backhoe, or the company that matched Julie's search for employment with a distant bank. In other words, we will need middlemen, brokers, publishers, and other intermediary people and organizations that will help us sort through the information.

This brings us squarely up against another popular myth: the arrival of the new world of information, we are told, will bring a new form of frictionless market, where people will buy directly from one another and will eliminate the middlemen. To be sure, this will happen if the costs of the middlemen are comparable to or higher than the costs of doing exactly the same thing in the Information Marketplace without them—such as transporting e-mail or buying a product you know you want from a seller you know carries it. But simply connecting producers and consumers of information and information services to one another via the Information Marketplace does not eliminate the brokers, any more than putting all the world's sellers and buyers in one location would eliminate the intermediate wholesalers. The pandemonium would be unprecedented. We will be inundated with so much junk that we will be unable or unwilling to devote our lives to sorting through it all. We will place great value on the intermediaries who can do this awful chore for us. And in turning to these intermediaries, we will notice that they have other traits we value, like their reputation and their ability to answer questions, which we cannot easily duplicate in the direct purchase scenarios. So before we plunge into direct, unbrokered transactions, we should exercise care in comparing all the visible and hidden costs (and benefits) of brokered and unbrokered transactions.

The same situation holds for the free exchange of information. Though it is egalitarian, even noble, to have a way for everybody to write about everything and "publish" it freely in the Information Marketplace, the result is the same: a huge pile of info-junk that most of us will not care about or read. Here, too, we will need the brokers of the written word, the visual and performing arts, and the new creative forms that will arise. Will software agents take care of this? I doubt it, because they won't be sufficiently intelligent for the task. More likely we will turn to the flesh-and-blood editors, publishers, and critics who can bring some judgment to bear in sorting the jewels from the info-junk.

So middlemen will not vanish. On the contrary, all sorts of them will thrive. The most valued will be those who attend to the segments of the Information Marketplace with the highest degree of confusion and the greatest potential for customization. Today's search engines on the Web would be an excellent place to start. People would readily pay for a finder service that is truly useful. Referral and matching services may be another, guiding you to services in the Information Marketplace that are known to be reliable and reputable, that can actually fulfill their claims, and that can match your needs.

The Rich-Poor Gap

We have focused this chapter on the value of information because so many social issues are rooted in economics and because we want to assess how they may be affected by the Information Marketplace. One of the most important—and most hotly debated—is its impact on the ancient clash between rich and poor nations and rich and poor people.

A wealthy nation has a rich basket of economic goods and services. It therefore values information and information services highly, because there are so many goods to which these lead. In the United States, Germany, and Japan, computer hardware and software and the information processing that goes on within organizations make up roughly 10 percent of the GNP. In Bangladesh the share is well below 0.1 percent—100 times smaller.

Similarly, if you are a rich person with a lot of goods and services at your disposal, then the intermediate information and information processing can go a long way toward helping you achieve your economic goals, because there is so much you have, and could have, to which information can lead. You are likely to value this information highly and pay for it handsomely. If you are very poor, then you have few goods, and there is very little of value to which information can lead. Therefore you place little if any value on information.

This inequity of information's value for rich and poor gives rise to an unfortunate instability. With the productivity gains made possible by all the information and information tools at their disposal, the rich nations and rich people of the world will improve and expand their economic goods and services, thereby getting richer. As they get richer they will leverage the Information Marketplace even further, thereby experiencing exponentially escalating economic growth. The poor nations and poor people, by contrast, can't even get started. They will tend to underuse information resources, because they can't afford them. They will gain no such leverage. There will be no rising spiral. They will stand still, which in relative terms means falling exponentially further behind the rich.

The painful conclusion is that, *left to its own devices, the Information Marketplace will increase the gap between rich and poor countries and between rich and poor people.*

Some thinkers counter that information technology can improve education and health, thus accelerating economic development in poor nations and unlocking poor people from their poverty. They go on to say that the poor may also deliver intermediate office work to the rich, thereby increasing their revenue and pulling themselves up the economic ladder. It is true that information technologies can be used to help people learn how to read, cultivate the land, generate electricity, and avoid illnesses. It is also true that machines may substitute for teachers in some training and educational tasks. It is certainly technically feasible for people to deliver intermediate office work at a distance. And there will even be an increased demand for such remote work as the world's firms become increasingly international and the Information Marketplace levels geographical differences. These are indeed exciting capabilities of the Information Marketplace that can help the poor. And they will come into play.

However, the hardware and software needed to achieve these lofty aims cost a great deal of money, as do the people who must accompany the equipment to instruct and aid neophytes in its use. This brings us squarely back to the key phrase in our conclusion above—that, *left to its own devices,* the Information Marketplace will increase the gap between rich and poor. Poor nations and poor people acting alone will not find the money needed to begin leveraging the Information Marketplace.

People sometimes suggest giving the poor an entrée to the Information Marketplace by making cheaper computers, perhaps with reduced functionality. This is unrealistic: With the cost of a factory to make integrated circuits approaching a billion dollars, the cost of the microprocessor chips that the factory churns out is independent of their function. It is driven primarily by amortizing high capital costs. And even if a microprocessor could be made more cheaply, it is not that big a component in an overall computer system. The screens, cabinets, and input-output devices account for a substantial portion of the cost. These are essential items that cannot be cut back appreciably. So though "reduced" computers might be made somewhat cheaper, even half as expensive, the end result—and our conclusions—would not change.

In many developing countries there is an even greater problem: the lack of a basic infrastructure to support computers, even cheap ones. It is sobering to observe that while the industrial West speaks of Internet connections to every home, only 2 percent of the homes owned by blacks in South Africa have a telephone. These countries need the more essential infrastructures of shelter, subsistence, and transportation before they can even consider an information infrastructure. The hope that people will jump from the agrarian to the information stage by skipping the industrial stage is as likely as a child's learning to dance without going through the walking stage!

Given human history, it is unlikely that the benefits of information technology will spontaneously accrue to the poor. Specific initiatives and programs must be started to mobilize people toward this worthy goal. The wealthy must help the poor access and use the new technologies. This can be done through various assistance programs, some already established. Kiosks in poor neighborhoods or in the African bush and public library–like computer centers in China and India will be much in demand and much appreciated, provided that they are supplied and staffed to take care of the many people who will flow toward them. Without such attention, the rich-poor gap will increase, with unpleasant, even dangerous consequences for those on both sides of the gap. As in prior social up-heavals that have widened class gaps, violent protests could well emerge as large groups of have-nots perceive themselves as steadily and hopelessly losing ground to the rest of the world.

These struggles would be no different and perhaps not as violent as clashes over goods essential to survival, like agricultural land and water. And yet the information case stands apart from factors that have aggravated class struggles in the past because it is so pervasive. There is hardly an activity that isn't or can't be touched and often improved by the effective use of information. The broad leverage of information that makes it so attractive to wealthy nations and wealthy people is also what makes it so devastating in its absence for the poor. We must help ensure that with respect to this critical gap the Information Marketplace is not "left to its own devices."

Employment or Unemployment

A big driver of a nation's wealth, and the standard of living of its people, is its ability to provide jobs. How might the Information Marketplace affect employment?

The politicians of the industrially wealthy nations are set on stimulating the Information Marketplaces of their countries, because they correctly perceive the economic leverage this move will bring. In their enthusiasm, they routinely affirm that the Information Age will not increase unemployment, because it will create many new businesses, hence more jobs. That position is to be expected, but is it realistic?

Employment in a line of business or across all businesses is the result of a race between productivity and demand. If the productivity of a nation increases by 10 percent and demand for its output also increases by the same amount in the same period, then there is no change in overall employment. But if productivity and demand don't move together, then employment will either increase or decrease. To understand how employment will actually change under the influence of the Information Marketplace, we need to understand how productivity and demand might change.

In the short term, the Information Marketplace will increase productivity only slightly, if at all, because we will still be learning how to make effective use of the new technologies. Nor is it likely to affect demand in any big way; people will buy a few more computers and software, but that is hardly a great change in the overall demand for goods and services. So in the near future, the Information Marketplace won't create any compelling pressure on the employment rate.

Some readers may find this conclusion hard to swallow. The leaders of the downsizing and reengineering movements of the 1990s have striven to justify massive layoffs as inevitable consequences of what they call the strategic uses of information technology. This is pure misdirection. To this observer and many others, these fashionable movements started in earnest to help a few organizations improve their performance by focusing on their processes and reducing unnecessary expenses. But these activities soon turned into convenient excuses for boosting short-term profits. At its absurd extreme, the method works like this: If you fire everybody today, then for a short time you will have immense financial success, because there will still be ample revenue coming in from prior sales with hardly any expenses. During that short period, whoever instigated the downsizing will be judged as a tough-minded miracle worker. Never mind that after a while the company will have to shut down. By then the miracle worker will be somewhere else "saving" another company!

The actual effects of downsizing and reengineering were, of course, more subtle. In many cases perhaps 30 percent of the people were laid off, and the company's prospective demise was neither quick nor obvious. When major companies around the world shut down their research capabilities, as they did on an unprecedented scale in the early 1990s, hardly anyone noticed. This indifference will continue until it becomes clear that these great companies have suddenly run out of ideas compared with competitors whose knives did not cut so deeply. Then, everyone will reach for the alarm bell, righteously indignant about massive corporate irresponsibility. In response, a new breed of consultants and gurus will magically appear with new schemes bearing catchy titles like BBB—Back to Business Basics. In due time, after much costly training of the new hires and a predictable reduction in profits, the movements will bring the companies close to where they were in the first place.

Is this view the frustrated raving of someone who loves research and hates to see it shut down? No doubt! But please consider this: had these reengineering movements happened half a century ago, they would have eliminated the positions and the people who invented the transistor at AT&T, the personal computer at Xerox, the computer network at BBN, and a host of other pivotal technological inventions that have defined progress during the last fifty years.

The unemployment caused by downsizing and reengineering is more the consequence of a preoccupation with short-term profits than of the strategic uses of information technology, and it should therefore not be credited to the latter. But enough about the short term. Over the long term, as we will see in the following chapter, the Information Marketplace will contribute to increased human productivity.

But what about long-term demand? Will it also increase? Here we are on virgin territory. To optimists, the answer is *definitely yes*. But this is less an educated guess and more a statement of faith. We will not be able to answer this question with any confidence until we can assess some of the early experiences of the real Information Marketplace and translate them to credible calculations on how productivity and demand are likely to grow, sector by sector.

We are therefore ignorant on this major issue. What we can predict, however, is that regardless of what happens with demand, a host of new jobs will appear on the scene, as was the case with the Industrial Revolution, which shrank the ranks of farmers and craftsmen but built the cadres of managers, engineers, pilots, and psychologists.

Taking our cues from our projections about new software tools and new uses of information, we can see how these new occupations might include all sorts of information service providers and telework brokers; conveners of groupwork; info-pilots who help navigate through the info-junk and info-tailors who customize the results; hyper-secretaries who can create accessible hyper-organized records of meetings; executives and experts who can network rapidly across the globe; info-artists skilled in the new media; authors and teachers who can create exciting hyper-materials; virtual travel agents; info-brokers; tele-consultants; marketeers for every conceivable informational or physical good and service and marketing specialists who deal in virtual markets; virtual pollsters who poll everyone on everything using the Information Marketplace; ratings specialists who can supply the success ratio of any service provider; electronic department store operators and employees; support professionals who help every conceivable profession use the Information Marketplace effectively; and so on. Whew.

Inevitably, even if demand tracks productivity perfectly, the Information Age will cause employment dislocations, because some of the people who provide the conventional forms of these jobs will be displaced. Prior dislocations have taught us that as we gaze with awe at the new job possibilities of the Information Age, we should also prepare ourselves to help individuals who will lose their jobs both tangibly and with compassion.

The Reach of Information

We will deal with the extreme case of reduced employment in the next chapter when we discuss the prospects for a work-free society. Meanwhile, let's wrap up our economic discussion by trying to identify the economic reach of the Information Marketplace. We'll start by asking what fraction of the economy it might affect.

Based on our discussion of the value of information, we can identify the portion of the overall gross national product (GNP) that deals with information and information work. It is the part of the GNP that includes all the final and intermediate information components. We'll baptize it the "gross national information product" (GNIP). As we have repeatedly noted, in the United States the GNIP stands at about 60 percent of the GNP.

What about the pervasiveness of information? Which of our activities might it touch in however small or big a way? Try to find an economic activity that is not affected by information. It seems impossible. Just knowing something about the activity, even that it is or is not taking place, is information. And there is someone, most likely the agent, beneficiary, or adversary of that activity, for whom this information has some value.

We conclude that *the Information Marketplace will touch essentially all human activity.* This is not to say that its impact will be uniformly significant or beneficial across this vast playing field—just that it could be very broad. And as we have seen, it could also be very deep in sectors and activities where automatization, group-work, and the other new tools we discussed are potentially useful.

As obvious as this huge reach is to us, most people are not aware of it—not even the people who are driving the big corporate mergers of the Information Age. I chaired a panel session on the Information Marketplace at a recent major international conference. One of the panelists was a famous media mogul who extolled the impact of the coming technologies on television but could not see any significant effects beyond the realm of entertainment. When I confronted him about his expectations for the Information Marketplace in commerce, health care, finance, education, government, law, and the many other areas we've discussed in this book, he dismissed them all by saying, "Blah, a bunch of idealistic stuff!"

Later that day we held a press conference. There were fifty journalists trampling one another to get at the media mogul. A fellow panelist, the dynamic CEO of a major computer company, and I tried to get him and the frenzied journalists to wake up and realize that what they were so excited about represented less than 5 percent of the economy. I cried out, "Some 95 percent of the economy out there is changing like wildfire under the impact of the Information Age, and you good people are oblivious to it all!" What I really wanted to shout was an equivalent and far more colorful Greek saying: "The world is burning and you are combing your pubic hair!" The entire boisterous gang ignored us!

When the moguls who, in theory, are at the vanguard of the Information Revolution don't appreciate its true potential impact, and the journalists who cover these moguls don't get it either, then the public can't help but have a skewed view. And every business will follow suit, waiting to see what the public will want to do.

For this book to do its service, I must now shed politeness and give blatant warning that this view is narrow, wrong, and possibly harmful. A company that ignores the full reach and impact of the Information Marketplace is planning its future on willful ignorance. It will view as irrelevant new technologies that might improve its business, like the automatization of office procedures; groupwork for task force deployment; data sockets and e-forms for sharing information with employees, customers, and suppliers; recruitment of employees over the nets—in other words, most of the tools (and associated benefits) discussed in this book. In doing so, it will undermine its efficiency, miss early business opportunities made possible by electronic commerce, and in the long run weaken its competitive position.

Companies that embrace the new world of information, by contrast, will gain valuable experience doing these things today, internally on their intranets and externally on the Web. When a full-fledged Information Marketplace is in place, they will be ahead of competitors that did not take the early plunge. Collectively, too, if a lot of companies ignore the bigger picture, they are likely to waste money and effort, prolonging the time it takes for the full Information Marketplace to mature and thereby delaying the much bigger profits all these players are after in the first place!

Sooner or later these skewed views will be history. At that point, what might the Information Marketplace's impact be on how we produce goods and services and how much of them we produce?

In some activities, like drafting, we can already measure 200 percent gains in worker output. In other areas, like word processing, measurements are fuzzy and results unclear, especially because it's difficult to define human productivity. Throughout this book we have discussed specific situations, for example, the use of hyper-organized summaries of meetings, in which productivity gains from the Information Marketplace might exceed 1,000 percent. These are interesting anecdotes but not enough to help us project overall economic impact. For that we must wait until more of the Information Marketplace is in place and has begun to show how it will affect each economic activity in each economic sector. To assess the overall economic impact on health care, for example, we must first experience the realities of having workstations at every medical person's station, kiosks in remote locations, a panoply of middleware and applications software, and human procedures that will dovetail all these functions together.

Could our information work be leveraged by the Information Marketplace by as much as our physical output was raised during the Industrial Revolution? The productivity gains contributed by the most recent wave of that movement, from 1890 to 1960, tripled output per worker in the United States. I suspect we might see similar gains, because of the parallels between the Industrial Revolution's effect on physical work and the Information Marketplace's potential impact on information work. But we cannot substantiate that suspicion yet. Might there be no change at all? In principle, yes. But in practice, I doubt it. With all that the Information Marketplace will touch, with all the candidates for improvement that we have discussed, and with what we already see taking shape around us, marked improvements in productivity seem very likely.

Could the balance between the huge amount of intermediate information and the smaller amount of final information change? Indeed. At one extreme, final information may dominate as people relieved of work by machines spend more of their time on leisure. At the other extreme, intermediate information is likely to grow because of the great value it derives from leading to almost the entire economy. The information component of a product like a pair of sneakers may thus grow, perhaps even dramatically, as companies use increasingly more information to leverage all aspects of their businesses. Across the whole economy, this would be translated to a growth pressure on the GNIP. But as the Information Marketplace improves office worker productivity, it will offset or even reverse this trend, because fewer information workers will be needed to achieve the same work. Eventually, if the GNIP follows the history of agriculture and manufacturing, it may signal a gradual yet substantial overall decrease in the number of human workers who deal with information.

Before we close this discussion, let's put the value of information in a broader human perspective. Information is not as valuable as food or any of the basic necessities of life. In a mortal battle of bytes against bites, people will always select food over data. And even if the information is valuable in that it may be critical to finding food, it can never be as valuable as the real food—by our argument, it will always be valued below the goods to which it leads!

Where does this leave us? Using our approach to valuing information, we have concluded with some certainty that the Information Marketplace will touch a huge part of our activities and influence about half of the industrial nations' economic might—although we don't know by how much. It is also certain that we will be inundated with mountains of info-junk and will continue to need intermediaries to help us find the nuggets we need. Human productivity will increase, but we cannot estimate what will happen to overall demand. Thus, although we can see over the horizon a wealth of new jobs in and around the Information Marketplace, we cannot anticipate what might happen to overall employment. Finally, the value of information leads us to a more certain, and less optimistic, conclusion about the poorer nations and people of the world. Left to its own devices, the Information Marketplace will exacerbate the rich-poor gap. Wealthy nations (and people) will need concerted strategies to address this undesirable and potentially dangerous situation.

We have made the tough journey—assessing some of the economic consequences of the changing value of information. Now it is time to marvel at the big societal changes that might result from the two major new forces that the Information Marketplace will bring upon the world—electronic bulldozers and electronic proximity.

12
Electronic Bulldozers

Plows, Engines, Networks

It is a few days before Christmas. I am out shopping at a well-known upscale department store in the Greater Boston area. I take nine items to the cash register. The cashier passes her magic wand over each package to read the bar code, and the impact printer rattles away as it prints a description and price for each item. I am getting ready to pull out my credit card when the woman turns to the cash register beside her and, horror of horrors, starts keying in the exact same information manually, reading the numbers off each package in turn. She is on package number 6 when I clear my throat conspicuously and, with the indignation of a time-study specialist, ask her why in the world she is duplicating the work of the bar-code reader. She waves me to silence with the authority of one accustomed to doing so. "Please, I have to finish this," she says politely. I tell her to take her time, even though my muscles are tightening up and my brain is engaging in vivid daydreams of punitive acts.

She finishes the last package, ignores my pointed sigh, reaches for a pencil, and starts all over again! This time she is writing in longhand on the store's copy of the receipt a string of numbers for every package. I am so shocked by this triple travesty that I forget my anger and ask her in true wonder what she is doing. Once more she waves me to silence so she can concentrate. Then she obliges: "I have to enter every part number by hand," she says. "Why?" I ask, with a discernible trembling in my voice. "Because my manager told me to," she replies, barely suppressing the urge to finish her sentence with the universal suffix *stupid!*

I could not let this go. I called for the manager. He looked at me knowingly and said with a sigh, "Computers, you know." I told him that this looked a bit more serious than that, and he proceeded to explain in slow, deliberate phrasing that the central machine didn't work, so a duplicate had to be entered by hand. "Then, why enter it at all by machine?" I ventured hopefully. "Because it is our standard operating procedure, and when the central machine comes back, we should be in a position to adjust our records for inventory changes." Hmm. "So why in the world is she also keying the numbers in, then?" I countered. "Oh. That's the general manager's instruction. He is concerned about our computer problems and wants to be able to verify and cross-check all the departmental entries."

I quietly walked out, stunned.

After I got over my shock at the absurd waste of time this store's procedures caused for the cashier—and me!—I began to marvel at how the great promise of computers to improve human productivity is more easily discussed than implemented. Technology detractors will say, "See, computers don't help us!" But they do, incredibly. Despite the temporary inconvenience, the store's computer system usually does make ringing up prices faster. More than that, it enables the store to automatically track inventory, which makes purchasing more efficient. And when prices change, as they frequently do, a simple keystroke for each item takes care of it instead of a stockboy with a sticker gun who has to relabel five hundred identical products. But clearly, as this incident shows, if we don't use technology wisely, it can make us less productive instead of more so.

Productivity is the yardstick by which socioeconomic revolutions are measured. Plows triggered the agrarian revolution by greatly improving the productivity of farmers. Engines, and later electricity, triggered the industrial revolutions by greatly improving the productivity of workers in manufacturing and transportation. If there is to be a true information revolution, then computers will have to repeat the pattern with information and information work.

Information technology has barely begun to improve productivity, and it has even hurt it in some cases; it takes longer to wade through those endless automated phone answering menus that it does to talk to a human operator. However, as I suggested in the previous chapter, productivity will rise once computers and communications are used in the Information Marketplace to relieve people of brain work in the way that industrial machinery relieved us of physical work.

Indeed, the Information Marketplace will give rise to two great new forces that will drive change in the twenty-first century: "electronic bulldozers" and "electronic proximity." This chapter deals with the first force. Chapter 13 deals with the second.

Ultimately, most of the hardware and communications technologies, human-machine interfaces, middleware, and information infrastructures we have discussed will either serve as electronic bulldozers or create electronic proximity. The bulldozers will relieve us of the burden of human work, either by completely replacing information-related human activities or by augmenting our ability to carry out these activities with less human work—in short, by increasing our productivity.

The world has largely ignored the cause-and-effect relationship between electronic bulldozers and productivity. Most people and companies buy new computers because the hardware has faster processors (more megahertz) or more memory (more megabytes or gigabytes), or because it is fashionable to own a new model, or because competitors have bought them and "we can't afford to fall behind." Imagine the absurdity of a company buying a new bulldozer because the motor turns at a higher speed, or because it's in vogue to do so, or because the competition just bought that model regardless of whether the machine can move any more earth in an hour!

Two hundred years ago, more than half of all Americans used their shovels, axes, and later their animal-drawn plows to produce the food that fed the nation. Today, about 3 percent of the people work toward the same goal. That's about a 20-to-1 increase in human productivity in agriculture, since we also produce more food today for a larger population than we did then.

The Industrial Revolution came along and turned all the displaced farmers loose on the engines, factories, transportation systems, and auxiliary professions that the Industrial Age created. The new kid on the block, manufacturing, became the big winner and rose to the same magical figure; by the end of the nineteenth century about half of the population in industrialized nations worked to produce the goods consumed by the citizenry.

Since then, the manufacturing establishments have progressively learned how to make goods with less and less human effort. Today, only 17 percent of the workforce in the United States is needed to manufacture all the goods that used to be made, plus many new ones—less imports, of course—another huge productivity increase. We are naturally led to ask, "Are we going to reach a stage where only 3 percent of the people will be able to manufacture everything we need?" We'll get to that topic shortly in our discussion of the prospects for a work-free society. For now, let's simply accept that productivity increases in manufacturing will continue.

So where have all the displaced people gone? To service jobs, which have been growing steadily. The key point to realize, however, is that a significant and growing portion of the service jobs, and the remaining manufacturing jobs, involve information. As we saw in the last chapter, the information-dominant economic activity, the GNIP, is now up to about half of the GNP in the world's industrial economies.

That's roughly the same portion of the workforce that was dedicated to agriculture and to manufacturing when those waves reached their peaks. If these magic patterns hold, it may be time for human productivity to begin its next surge, this time through electronic bulldozers. The social patterns are there in our own lives to remind us of this progression, too. Most of our great-grandfathers worked the land; our grandfathers shed plows for assembly lines; our fathers and mothers moved to the office; and now, we and our children are beginning to use the Information Marketplace on an increasingly regular basis.

Productivity will rise in the Information Age as it did in the Industrial Age and for the same reasons it did before: the application of new tools to relieve human work. To ignore computers' fundamental ability to help humans do their brain work is at best perverse and at worst irresponsible.

Let's explore how the Information Marketplace might help us in the eternal quest to get more results for less work. To do this, we will first examine a series of "faults"—ways in which computer technology is misused today, because of either technological or human foibles. Correcting these faults will be the first step toward increasing our productivity. Making the Information Marketplace easier to use will be the second step.

What's Wrong with Technology

The additive fault. The ridiculous duplication of effort that I ran into at the department store happens often and in many different settings. We'll call this failure the additive fault, because in these cases people are doing everything they used to do before computers, plus the additional work required to keep the computers happy or to make the people appear modern. In anybody's book, this is a mindless productivity decrease! It should be stopped cold in whatever setting it raises its ugly head. And while we are at it, let's recognize this particular problem for what it is: It's not caused by technology. It's caused by our own misuse of technology.

The ratchet fault. Sometime after my encounter with the cashier, the same advance gremlins that seem to run ahead of me to set up challenging situations must have surely visited the airline clerk at Boston's Logan Airport. When I arrived I handed the clerk my ticket to New York and asked him to replace it with one to Washington, D.C. "Certainly, sir," he said, and bowed to his terminal, as if to a god. As a seasoned observer of this ritual, I started recording his interactions. Bursts of keystrokes were followed by pensive looks, occasionally bordering on consternation, as hand-on-chin he gazed motionless at the screen, trying to decide what to type next. A full 146 keystrokes later, grouped into twelve assaults demarcated by the Enter key, and after a grand total of fourteen minutes, I received my new ticket.

What makes this story interesting from a productivity perspective is that any computer science undergraduate can design a system that does this job in fourteen seconds! You simply shove your old ticket into the slot, where all its contents are read by the machine. You then type or speak the "change" command and the new destination, and you get the revised ticket printed and shoved back in your hand. Because fourteen minutes is sixty times longer than fourteen seconds, the human productivity improvement with such a box would be 60 to 1, or 6,000 percent!

Something is terribly wrong here. People run to buy a new computer because it is 20 percent faster than the one they have, and we are talking here about a 6,000 percent improvement. So why aren't the airlines stampeding to build this box? For one thing, if they did this for every one of the possible requests, they would have to build a few thousand boxes. All right then, why don't they reprogram their central computers to do this faster? Because that would cost over a billion dollars. Why? Because they have been adding many software upgrades and changes to their systems every year. After twenty years they have built up a spaghetti-like mess that even they cannot untangle. In effect, they cannot improve their system without starting from scratch.

We'll call this the ratchet fault of computer use because it's like a ratcheting tire jack: every time a new software modification is added the car goes up, but it never comes down unless a precipitous event, like a total redesign, takes place. This problem is more a problem of technology than of human practice. If we had a software technology that could let us gracefully update our systems to suit our changing needs while maintaining their efficiency, then we wouldn't be in this bind.

The excessive learning fault. One-tenth of my bookshelf is occupied by word processing manuals. Add the manuals for spreadsheets, presentations, and databases, and I'm comfortably up to half a shelf. Because I use graphics and do a bit of programming, I need a few more manuals. This brings the total length of my computer guidebooks to one EB—one (printed) *Encyclopaedia Britannica*. We'll simply call this the excessive learning fault—the expectation that people will learn and retain an amount of knowledge much greater than the benefits they'd get from using that knowledge. Imagine requiring people to digest an 850-page manual in order to operate a pencil. We laugh at the thought, but we accept it readily in the case of a word processing program!

Though there are all kinds of rationalizations about the difference between human expectations and machine capabilities, we cannot get away from the fundamental problem that people today are expected to learn many unnatural and arcane computer commands just so they can harness their machines. I have little doubt that the first half of the twenty-first century will be spent getting rid of fat manuals and making computers much easier and more natural to use. True ease of use is central to the quest for greater productivity, and we'll devote the following section of this chapter to it.

The feature overload fault. This next one is intimately related to the excessive learning fault. *Bloated* is perhaps a more accurate adjective to describe the feature-packed programs hitting the market in the late-1990s. Vendors do so in part to cover their bets and to be able to charge higher average prices. Buyers are fascinated by the potential uses of their computers and value their prerogative to command their machines to do thousands of different things. Of course, in practice they end up doing only a few tasks and forget what features they have bought or how to use them. Yet the demand for features is so strong that software vendors stuff programs with many megabytes of unnecessary appendages, which lull buyers into tolerating delays, crashes, and thick manuals—all caused by the added features. A top-selling "suite" of office software comes on a CD-ROM or forty-six diskettes that require half a day to load into your machine! This is not productive. And it is caused by us, not technological weaknesses. Consumers and corporate executives should declare birth control on the overpopulation of excessive and often useless features.

The perfection fault. None of us is immune to this next activity. When we create documents, we spend a lot of time adjusting margins, changing fonts and styles, choosing different colors, and generally fussing over the appearance of our information—as if the message "You will get a 5 percent raise" or "Your position has been eliminated" would be substantially improved by appearance. It is counterproductive for an author to create a very pretty letter, an incredibly beautiful spreadsheet, or a gorgeous overhead transparency when a simpler one that conveys exactly the same information and takes half the time to compose will suffice. I suspect that in a social movement akin to reverse snobbery, people will soon revert to plainly clothed information output, perhaps elevating the practice to a status symbol that signals a greater care for content than for form. Of course, a good appearance does count for something. So people will eventually settle for whatever balance between aesthetics and utility they find comfortable.

The fake intelligence fault. My car has a fancy phone that was advertised as "intelligent" because when it makes a phone connection it automatically mutes the volume of the car radio to ensure a quiet environment. I found it a delightful feature until one afternoon when I heard a good friend being interviewed on the radio. I immediately called a mutual friend so she could listen along with me over the phone and share in the excitement. This, of course, was impossible, because the phone muted the radio and I couldn't override it. Welcome to the fake intelligence fault. It crops up in many situations where a well meaning programmer puts what she believes is powerful intelligence in her program to make life easier for the user. Unfortunately, when that intelligence is too little for the task at hand, as is always the case, the feature gets in your way.

Faced with a choice between this kind of half-smart system and a machine with massive but unpretentious stupidity, I would opt for the latter, because at least then I would know what it could and couldn't do. So before you agree to use (or buy!) an intelligent agent or a knowbot or some other purportedly knowledgeable program, you'd better check it out. It will almost always suffer in some way from this fault, because we do not know yet how to make programs with broad cognitive capabilities, common sense, and other attributes that would let their intelligence track ours. As users striving to improve our productivity, we must always ask of a new program, Does it offer enough value through its purported intelligence to offset the headaches that it will inadvertently bring about? If so, then go for it. And the first imperative before us as suppliers of these ambitious programs should be to provide users with a Go Stupid command that lets them disable the intelligent features!

The machine-in-charge fault. It is 2:00 A.M. and I just got home. My Swissair flight from Logan was canceled because of trouble in the motor controlling the wing flaps. Some 350 passengers whose plans were thwarted were bombarding every available clerk at the airport. I abandoned that zoo, rushed home, switched on my computer, and tried to connect to the Easy Sabre do-it-yourself airline reservation service offered by Prodigy to search for an alternative ticket for a morning flight out of either Boston or New York. I had to find out before going to sleep if this was possible. I logged on, but before I had a chance to enter a single keystroke, Prodigy seized control of my screen and keyboard! It informed me that to improve my system's use of its online services, it would take a few moments (meaning, a half-hour minimum) to download some improved software.

There was nothing I could do to stop Prodigy from "helping me" in its own murderous way. A meager piece of anonymous software was in full control of this situation, while I, a learned human being, was pinned against the wall, destined to wait for half an hour. Meanwhile, with each minute that passed, another of those frantic nomads at the airport would take another of the rapidly vanishing seats on the next morning's few flights. I gladly would have used software that was several generations old to get my job done sooner. I felt I was drowning in the shallow surf from a stomach cramp while the lifeguard on the beach was oblivious to my screams because he was using his megaphone to inform me and all the other swimmers of improved safety procedures.

This is exactly the same fault we discussed early on when we observed that distinguished humans were executing machine-level instructions dispensed by hundred-dollar automated telephone operators, with their familiar "If you want Marketing, please press 1. If you want Engineering..." A good part of this machine-in-charge fault must be attributed to human failure in allowing such practices to continue without objection. Programmers must also take some of the blame. They often deliberately use this fault, because it's simpler, therefore cheaper, to program a computer to interrogate the user and not let go until all questions have been answered in one of a few fixed ways than it is to allow the user to do any one of several things with the assurance that the computer will pay attention.

As we have already discussed, interactions like these are not always undesirable. A mistaken command by you to erase everything inside your computer should not be casually executed. However, 95 percent of the overcontrolling interactions on the world's computers don't involve such grave situations. They are widely used because they increase the programmer's productivity, even though they *de*crease user productivity! The sooner these software crutches vanish and the user is given control, the sooner machines will serve humans rather than the other way around.

The excessive complexity fault. We'll conclude with a type of fault that makes an even more pressing point about the productive use of future information systems.

I am at my office, it is almost noon, and I discover with considerable panic that I forgot in my home computer the crucial overheads I need for an imminent lunch meeting. No sweat. I'll call home and have them shipped electronically to my office. As luck would have it, though, the only one home is the electrician. However, he is game and agrees to do exactly what I tell him. "Please turn the computer on by pushing the button on top of the keyboard," I say. He is obviously a good man, because I hear the familiar chime through the phone. Two minutes later the machine has booted up. While this is happening, the electrician wants to know why the machine doesn't come on instantly, like a light bulb.

I refrain from telling him that I share his consternation. For three years I had been trying to interest sponsors and researchers in a project that would address this annoying on-off business in which a human respectfully begs permission from a computer's software to turn it on or off. Instead, I explain that the machine is like an empty shell and must first fill itself with all the software it needs to become useful, ". . . and that, my friend, takes time. Okay, pull down the Apple menu and select the Call Office command," which I had providentially defined some time back. He complies, and I hear my home modem beeping as it dials my office modem. On the second ring I hear the office modem next to me answer. We are almost there, I muse hopefully. "Do you see the message that we are connected?" I ask as I explain what he should see. "Nope," he responds. Another minute goes by, and he reads me an alert message that has appeared on my home computer's screen. I know what happened. The modems latched correctly, but the software of the two machines did not. I ask him to hold while I restart my machine. Like many people, and all computer professionals, I know that restarting with a clean slate often solves problems like this one, even though I have no idea what actually caused the problem.

As I guide the electrician through a rebooting I get angry, because these problems would be reduced if my office computer were calling my home machine rather than the other way around. But my home machine has only the so-called remote client software, meaning that it can only call out. So the electrician has to have my home computer call me. Another monumental stupidity of current technology, fostered by the fashionable and ill-advised trend toward dividing the world's computers into either "servers" or "clients." Imagine approaching someone to ask a question and being told that he is only a client and cannot dish out information to you. This asymmetry is a residue of corporate computing and the time-shared era's central machines that dispensed lots of data to the dumber terminals. The distinction has no alternative but to vanish so that all computers (which I'd coin "clervers") will be able to dish out and accept information equally, as they must if they are going to support the distributed buying, selling, and free exchange of information that the Information Marketplace is all about.

My home machine has again booted up. We go through the modem dance once more, and this time the software latches. I ask the electrician to select the Chooser command and click on the Appleshare icon. He does so. We are almost there. I ask him to click on the image of my office machine. Now he needs my password, which I give him promptly . . . and I won't change it later. He reports activity on his screen that I interpret as success. I tell him how to locate the precious file I need and send it to me. In two and a half more minutes the overhead images are safely in my machine. I thank the electrician profusely and send the images to my printer, now filled with blank transparency sheets, and I've got them. I arrive at the meeting thirty minutes late.

This final fault in our litany, the excessive complexity fault, is attributed entirely to technologists. It is inexcusable at the dawn of the twenty-first century to design systems for human beings that bring this much complexity and hassle to the simplest of tasks—in this case, sending some information from here to there.

Why couldn't I simply give my home computer in one second a single command like "Send the overheads I created last night to my office" and have them arrive three minutes later? Techies, please don't tell me it can be done with a different kind of machine or a different operating system, macros, agents, or any other such tools, because I know and you know better. Indeed, this simple act just can't be carried out easily and reliably with today's computers.

Technologists must begin the long overdue corrective actions against the complexity fault by simplifying options, restricting them, and, most important, *reversing* a design point of view rooted in decades-old habits. They should tailor computer commands and options to user's needs, rather than tailoring them to existing system and subsystem needs and expecting users to obediently adapt. As users we shouldn't have to tell the modem or the network what to do any more than we might need to fiddle with the fuel-air mixture or the timing of the ignition to drive our car. If we really want to improve our productivity, we must design software that truly orients the controls to a higher level, closer to the user's desires and goals—the equivalents of steering wheels and brakes—ensuring all along that the subsystems work harmoniously and reliably toward these goals, fixing any troubles that may develop along the way. Systems that purport to do this today break down too easily, revealing the underlying miasma and leaving us helpless. It is a challenging task worthy of the best technologists.

I'd like to note, too, that people everywhere who encounter these kinds of problems in their information tools should not feel that they are the ones causing them or that they could have avoided them if they had carefully read all the hundreds of pages of manuals. Rest assured that computer professionals have exactly the same frustrations and pains. The author is often reduced to crying on his knees as he tackles such frustrating situations. And when he calls in his expert associates for help, they too end on their knees crying!

My diatribe against computer faults might lead you to think I believe they are all the result of careless design and use. Not so! Viewed collectively, they are the early attempts of the ancient humans that we all are to digest and control an Information Marketplace that's growing faster than our ability to understand and accept it.

We also should not be too overcome by these difficulties. Sure, these problems are frustrating when we suddenly come up against them. And when we are in that mood, we are ready to turn our computers into boat anchors to avoid the pain. Yet after a problem has been fixed, we are back at our machines with starry eyes. The bumps in the road are understandable for a young and growing field. Remember those early airplane tests we see on movie reels, in which everything falls apart? In due time the bumps will be smoothed out, and we will be able to derive the productive power of the Information Marketplace for our purposes. Instead of lamenting about the faults, we should recognize them objectively, guard against them and any new ones that may arise, and focus on avoiding and reversing them. That's one sure way to make our electronic bulldozers move more earth.

Ease of Use Revisited

Fixing problems is the first step to greater leverage of the Information Marketplace. But making the technology easier to use is the really big lever.

In the last decade, anyone who has used the phrase *user friendly* in my presence has run the risk of physical assault. The phrase has been shamelessly invoked to suggest that a program is easy and natural to use, when this is rarely true. Typically, *user friendly* refers to a program with a WIMP interface, meaning it uses *w*indows and *i*cons, *m*enus, and *p*ointing along with an assortment of pretty colors and fonts that can be varied to suit users' tastes. This kind of overstatement is tantamount to dressing a chimpanzee in a green hospital gown and earnestly parading it as a surgeon! Let's try to penetrate this hype by painting a picture of where we really are today, and where the true potential for ease of use lies.

It is sometime in the late 1980s. A friend approaches you, excited by his ability to use spreadsheets. You ask him to explain how they work. He shows you a large grid. "If you put a bunch of numbers in one column," he says, "and then below them put the simple command that adds them up, you will see their total in the bottom cell. If you then change one of the numbers, the total will change automatically." The friend rushes on, barely able to control his exuberance: "And if you want to make the first number ten percent larger, you just put in the cell next to it the simple command that multiplies it by one-point-one." His expression becomes lustful: "Do you want to increase all the numbers by ten percent? Just drag your mouse down like this, and they will all obey." He takes in a deep breath, ready to explode once more, when you stop him cold. "Thank you. Now go away," you say. "You have taught me enough to do all my accounting chores."

This is how millions of people today use their spreadsheet programs like Microsoft Excel and Lotus 1-2-3. They hardly know more than a tenth of the commands, yet they get ample productivity gains. Let's move on.

You are happy with your newly acquired knowledge until one day you discover that you need to do something a bit more ambitious, like repeat over an entire page all the laborious operations you have set up but with a new set of initial numbers. Perplexed, you go back to your friend, who smiles knowingly and tells you that you must now learn about *macros*. His explanations are no longer as simple as before, and you just can't get the spreadsheet to do what you want. If you're like most of the millions who use spreadsheets, this is where you would give up. But instead you fight on, eventually mastering the mysteries of the macro. It's really a computer program written in an arcane programming language that replaces you in commanding the spreadsheet program to do things you would have done manually.

You sail along for the next six months, until you develop the need to do an even more ambitious task that involves controlling human-machine interfaces and more. You go back to your friend who tells you that you have become too good for the limited capabilities of this spreadsheet application, and that you must now learn how to use a real programming language like C++. Unaware of what lies behind these three innocent symbols, but unwilling to give up, you press on. This costs you your job, because you must now devote full time to a colossal new learning endeavor. Yet you are so enamored with programming that you don't mind. In fact, you like the idea. Two years later, having harnessed C++ and a few more programming languages and operating systems, you begin a career as a successful independent software vendor and eventually become very wealthy.

This happy ending cannot hide the barriers that you have had to overcome along the way. You decide to graph the effort you expended versus the capability you gained. The result is a line starting at the left and moving along to the right. There is a long slowly rising portion, and then a huge hill where you had to learn a lot of new stuff in order to move further right. Then there are more slowly rising lines, and more huge hills, like a mountain chain where each new mountain gets higher. You wish that someone would invent an approach with a gentler slope, one where you get increasingly greater returns as you increase your learning effort, without the impossible cliffs that you had to climb. I predict that such "gentle slope systems," as I like to call them, will appear and will mark an important turning point of the Information Age.

If by now you are sure that this story is irrelevant to your purposes because it involves the frightening notion of programming, I ask you to hold on a bit longer. I will soon demonstrate why each of us will find it necessary and desirable to follow such a route.

The gentle slope systems will have a few key properties. First and foremost, they will give incrementally more useful results for incrementally greater effort. They will be able to automate any activity you do that is repetitive. They will be graceful, in the sense that incomplete actions or errors on your part will result in reasonable degradations of performance rather than catastrophes. Finally, they will be easy to understand—no more complicated than reading a cookbook recipe.

Constructing such systems will be an important challenge for the information technologists of the twenty-first century. Some forerunners have begun appearing, with names like Powerbuilder and Appware. But full-fledged gentle slope systems are still a long way off.

Another reason it is difficult for nonprogrammers to tell computers what to do is that the software systems that surround us are preoccupied with the structure rather than the meaning of information. We can program them to do anything we want, but they are unaware of the meaning of even the simplest things we are trying to do. Let me illustrate.

It takes me seventeen seconds to say to a programmer,

Please write me a program that I can use to enter onto my computer the checks I write, along with the categories of each expenditure—food, recreation, and so forth. And do this so that I can ask for a report of the checks that I have written to date, listed chronologically or by category.

I have given this assignment several times to different people. Master programmers invariably decline to play and tell me to go buy this program because it's commercially available. Good programmers will say they can program it in a couple of hours . . . and end up taking a day or two to develop a shaky prototype by the time all the wrinkles have been smoothed out. Inexperienced programmers will say cockily that they can write the program in a few minutes as a spreadsheet macro . . . and are generally unable to deliver it at all. The company Intuit, which developed the very successful Quicken program that does this job and more, took two years and many millions of dollars to get the software developed, tested, documented, and brought to market.

Why can I "program" a human being to understand the above instruction in a mere seventeen seconds, while it takes a few thousand to a few million times longer to program a computer to understand the same thing?

The answer surely lies in the fact that humans share concepts like *check, category, report*, and *chronologically* while computers do not. The machine is so ignorant of these concepts that the programmer must spend virtually all of his programming time teaching the computer what they mean.

Here, for example, is how my thinking as a programmer would unfold in planning the coding for a very small part of this program: *I'll use tables to represent the checks. Each entry will have five fields: check number, to whom paid, amount, category of expenditure, and a blank field for something I am sure I'll need later. Okay, now when the user assigns a category to the check, I'd better have a list of acceptable categories to display in a pop-up menu, so he doesn't have to type in a category every time. This will also prevent misspelled entries. This means that I will need a mechanism for the user to enter, edit, and delete categories of interest to him. There is no such built-in mechanism in the system I am using to develop this code, so I'd better add to my list of programming tasks this necessary subsystem.*

Multiply this paragraph by a thousand, and you begin to get an idea of what it takes to program a real computer application.

If, however, I had a computer that already understood some of these "concepts," then I might be able to program it to do my job in a very short time. This is a deep and important way in which computers could increase our productivity in the twenty-first century: by being made to understand more human concepts in better ways.

It's unlikely that computers will achieve this understanding as well as people do anytime in the next century. As we have already noted, we have no clue how human common sense works, and we cannot teach computers to understand what we do not understand ourselves. Yet we need not despair. As we have seen with the use of e-forms, some less ambitious and simpler ways to handle shared concepts are possible and can be very powerful.

If we are to make computers truly easier to use, techies will have to shift their focus away from the twentieth-century preoccupation with the structures of information tools like databases, spreadsheets, editors, browsers, and languages. In their early stage, computers became ubiquitous because this focus allowed these common tools to be used equally in thousands of applications, from accounting to engineering to art. Yet that same generality is what makes them ignorant of the special uses that they must ultimately serve, hence makes them less useful—much like a dilettante jack-of-all-trades!

What we need now, to increase utility further, is a new breed of software systems—an accounting "spreadsheet" that an accountant can easily program and that already "understands" higher-level repetitive tasks like setting up charts of accounts, doing a cash reconciliation, and pulling trial balances; and then a stockroom "spreadsheet" that a warehouseman can easily program to rapidly perform tasks like readjusting inventory levels and rearranging orders by physical location so he can pick parts in a single, straight-line pass through the warehouse; and then a "spreadsheet" that knows about shipping for the shipping department; and so on. Freed from the tyranny of generality, these specialized "programming" environments will rise toward offering a lot more of the basic information and operations of their specialty. The time has come for computer technologists to abandon the "generalist" orientation that served people well for the first four decades of the computer era and *shift their focus from the structure to the meaning of information.*

By now you must be frustrated by my stubborn preoccupation with programming. Why not leave the programming to the programmers, as we do today, and let ordinary mortals buy the applications that these programmers develop? This may sound reasonable, but it is not—at least, not if we want to achieve the biggest promise of the Information Age: the great and still unrealized potential of tailoring information technology to individual human needs.

Today's applications programs are like ready-made clothes. One size fits all. So most of them are ill-fitting, and we have to contort ourselves to improve the fit. Another outcome of this practice for business is that, if every company used the same set of canned programs, they would follow more or less the same procedures, and no company would stand out against the competition. Shrink-wrapped, ready-made software is good enough for the state of information technology at the end of the twentieth century. But it won't be as good in tomorrow's Information Marketplace.

Great gains will be achieved when you can bend and fashion information tools that already understand a great deal about your specialty to do exactly what you want them to do, for you or your company, rather than bending yourself to what the tools can do. This quest for customizable electronic bulldozers with specialized knowledge will be no different than the current trend toward customizable manufacturing. It could well be that by the close of the twenty-first century, a new form of truly accessible programming will be the province of everyone and will be viewed like writing, which was once the province of the ancient scribes but eventually became accessible to everybody.

This isn't as absurd as it sounds. We invented writing so that we could communicate better with one another. Tomorrow we'll need to communicate better with our electronic assistants, so we'll extend our "club" to include them as well. Everyone will then be a "programmer," not just the privileged few. And none of them will be conscious of it.

In fact, this is already happening on a small scale with the millions of people who use spreadsheets and would be very surprised to learn that they are programmers. When I say people will program, I am not talking about writing the detailed code and instructions that make computers run. This will still be the bulk of a software program and will indeed be pre-done by professional programmers, who will fashion the numerous larger building blocks that we will use. Each individual's "programming" will account for a very small fraction of the software code, maybe 1 percent. But it will be the crucial factor that gives the program its specificity. It will be like building a model railroad; you don't make all the track or engines or cars, but you do arrange the pieces to create your own custom railway patterns.

Beyond correcting today's computer faults, greater productivity in the new world of information lies in making the Information Marketplace's tools easier to use. Techies can strive toward this noble goal by creating a new breed of software that have gentle slope learning curves, that understand the meaning of certain useful concepts in specialized areas of human activity, and that can be easily customized with little effort by ordinary people to meet their specific needs and purposes. These shifts, and greater automatization, will comprise the key that turns today's incomprehensible cacophony of computer grunts into a chorus of understanding and synergy among our electronic bulldozers as we harness them more effectively toward fulfilling our human desires.

What's the Horsepower of Your Text Editor?

Once all this happens, we will be on our way to a world where we will experience greater gain for less work. Before we consider the consequences of this promise, however, we have to know one thing: whether we are indeed becoming more productive. That means we have to figure out a way to measure the benefits.

Let's say your car has a 200-horsepower engine and costs $10,000. It can do the work of two hundred horses at a cost of $50 per horse. What is your computer's horsepower? Well, you know the cost, but how do you figure out how many horses or people or other workers it replaces?

If you are doing some form of computer-aided design (CAD), the answer is straightforward. A properly trained designer with a computer and the right software can do, on average, the work of three designers equipped with the older tools of straightedge, pencil, and compass. So we could rate a CAD system as leveraging the work of one person into 3.0 hdp (*h*uman *d*esigner *p*ower). And in due time we might replace it with an equally expensive 3.5 hdp system.

But what about someone navigating the Internet, using a word processor, or shopping via a speech interface? What about an accountant using a spreadsheet, or a teacher preparing lessons, or a travel agent booking a flight? What is the productivity increase they can achieve in the Information Marketplace?

There is no single answer (or single number). Computers in the Information Marketplace will help people with their work in at least three broad ways—"automatization," "augmentation," and "mediation." Let's consider how to measure the productivity gains for each.

As we've already seen, automatization involves the assumption of human information work by computer. The implementation of Dr. Kane's standing procedures, by which X rays, bills, and other paperwork are automatically routed and processed, is one example.

Another is the explosion of an order for a new car into all the subsystems needed to make up that car. In automatization, computers and communications liberate people from recurring and boring work, usually through preset standing procedures. It seems reasonable to expect that we will be able to measure the productivity of automatization tools through their cost and the equivalent (specialized) human work each system replaces.

Augmentation is the help a computer provides by amplifying a person's ability to do a specific task. Augmentation differs from automatization in that a human is still involved and controls the quality of the output. This is the traditional way in which standalone computers help single users. The word processor and the CAD system are examples. Here, it would seem reasonable to measure the productivity gain as the added amount (and kind) of work it enables one user to perform, for example, 3.0 hdp in the case of the CAD system. The important distinction is that qualitative comparisons must also be made, because the output involves the augmentation of human quality, which is difficult to measure.

Mediation is the use of computers and communications to help human-to-human exchanges. Groupwork, e-mail, bulletin boards, tele-medicine, discussion groups, and employment at a distance are all made possible through mediation. Here, the work done by the group with the computer is an improvement over the work done by the same group without machine help. In this century the telephone has been our primary mediator. The productivity gained through mediation is even harder to measure, because it mainly improves human-to-human activities like communicating and groupwork, which are almost entirely qualitative and hard to define in productivity terms. Ultimately, we will have to better understand what this important factor is. Perhaps we will do it comparatively by assessing quantitatively and qualitatively the achievements of groups that use mediation against like groups that do not. How do the firefighting teams of British Petroleum fare in cost, speed, motivation, and success rate versus the same teams before they started using their own groupwork technology? The answer may well be a paragraph of conclusions.

Greater Productivity

Now that we have discussed the different kinds of productivity that information technology's electronic bulldozers can bring, and the need to measure the gains, let's revisit the more fundamental question of whether the Information Marketplace will raise society's overall productivity and, if so, how quickly.

Believe it or not, we don't seem to understand the productivity of today's office workers, whom we expect computers are so eminently suited to augment. It's hard to determine quantitatively how the information work of a clerk or a chief executive differs. Though it is possible to measure the number of accounts payable forms the clerk processes in a day and the profits the CEO is accountable for at the end of the year, we still don't understand how to truly account for their overall productivity. Faced with this problem, economists often measure office productivity by subtracting manufacturing productivity from the overall productivity of the workforce. After making certain additional adjustments, they end up with the contribution to productivity of what is left over—office workers.

Unfortunately, this approach can lead to confusion. A widely publicized report by Stephen Roach, chief economist of Morgan Stanley, showed that while factory productivity grew by 17 percent during a seven-year period in the late 1980s, office productivity decreased by 7 percent. The report never attributed the decline to computers, and much subsequent explaining and backpedaling by all sorts of people added even more confusion to the claim. Roach's observation was a big surprise. It does not take rocket science to conclude that while the sale of computers grew by leaps and bounds to reach 10 percent of the U.S. economy, office work productivity did not skyrocket as expected.

Many explanations were offered for this *productivity paradox,* as it became known. They ranged from the adverse effects on productivity of a growing mountain of governmental regulations, to errors of measurement and the relative youth of the computer field. We may never know what gave rise to the productivity paradox. As some have suggested, it may even turn out not to have been a paradox. Never mind. In retrospect, its principal contribution may turn out to be that it brought within striking distance two words—*computers* and *productivity*—that my colleague Nobel economist Bob Solow claims never appear together. It is also likely that the true contribution of information technology to human productivity will not be known until a good part of the twenty-first century has elapsed and the world has discovered in some detail the ways in which computers help or hinder human quests.

Nevertheless, we can make some observations on computers and productivity based on how information technology has helped companies to date. Paul Strassmann, for many years the chief of information at Xerox and more recently director of information at the Pentagon, conducted an ambitious study of 630 different companies, asking precisely this question. In his book *The Business Value of Computers*, he attempted to relate each company's profitability to various indicators of computer intensity like the information technology budget and the number of PCs per employee. His many charts show no correlation at all—until he splits the companies into two groups, according to whether they are well managed (in his words, have a high return on management) or not. On average, the well-managed firms with high computer intensity do better than the well-managed firms without it. And intriguingly, the ill-managed firms with high computer intensity do worse than the ill-managed firms without it. This result, like all great results, surprises at first and then becomes obvious in retrospect. Information technology acts as a magnifying lens, amplifying management's strengths but also its weaknesses. A backhoe or a scalpel can do marvels in the hands of a skilled operator . . . or a great deal of damage in the hands of an impostor. The same goes with information technology.

The electronic bulldozers of the Information Marketplace will indeed help increase human productivity. The process is likely to be slow, with significant changes being felt over the long term, because it takes time for people to change their habits and procedures and to learn new ways of working. As we saw, this was also the case with the Industrial Revolution, which tripled factory worker output but took nearly a century to do so. As we have already discussed, we cannot prove that such a productivity increase will definitely happen, because the Information Marketplace and its electronic bulldozers are just beginning to be felt. But all the early indications before us and all the prospects that we have discussed give us confidence to be optimistic.

Toward a Work-Free Society

Once we accept the notion that productivity will increase significantly, we can then ask how this might affect society. Let's dare to leap a couple of centuries into the future and speculate about a new era that we'll call the "Work-Free Society." The barons and princes of this period will be the people who own large, fully automated factories and fully automated service dispensing centers that provide essentially all the goods and services needed by people. The less wealthy folk will make do with a handful of ma-and-pa-owned ma-chines that also produce, automatically, a handful of products and services at a smaller scale. On her deathbed, the middle-class citizen of the twenty-third century would will to her beloved son her pencil- and paper-making robots (let this be a bet that there will still be pencils and ten times as much paper).

From manufacturing to health care, the generation of the world's goods and services will be largely in the hands of machines, which will be the principal property owned by all people. The people of the world will do no work, because they will derive all the revenue they need to buy their desired goods and services from the machines that they own. Machines will make the machines that are needed, too.

Two requirements would have to be met for this wild scenario to materialize. On the humie side, people will have to espouse the work-free ethic. On the techie side, the Information Marketplace and a huge arsenal of physical machinery will have to ascend, dramatically, the ladder of technological progress so that they can generate all farming goods without farmers, all needed products without factory workers, and all office work without office workers.

At first blush, the humie requirement doesn't seem that difficult; we are always willing to get more for doing less. On second thought, however, this transformation may not be so easy. Another part of us would be confused. Roughly three-quarters of our economy is generated with human labor and only one-quarter with capital. This ratio has remained remarkably constant for nearly a century. Economists would have every right to be skeptical about its changing so radically, even over a couple of centuries. More important, at a philosophical and psychological level, we do not have a precedent or an ideology for a work-free society. We have witnessed tiny fractions of society that did not have to work, like the aristocracy of various periods, but we have not experienced a work-free environment across the board. Human beings have always had the obligation and later the right to their work, which they have exercised and continue to exercise. And the wish to earn more money, satisfaction, or glory through additional work has been a powerful driving force for rich and poor alike. We may therefore have mixed feelings about accepting a work-free society even if it were technically possible.

The techie requirement is somewhere between improbably difficult and impossible. Information machinery would have to improve dramatically in order to produce the world's goods and services without human help. There is simply too broad and too rich a spectrum of activities that human workers carry out that we do not understand how to do by machine. Motivating and inspiring top the list. Common sense is next. Way down the list, there is a myriad of apparently small tasks like ironing a shirt or cleaning the insides of your cabinets that by now we should have been able to automate but that remain completely out of the reach of our automated servants. Who would tend to surgical tasks, or teaching, driving the school bus, protecting us from crime, running our governments, piloting the aircraft we ride, investing our money, representing us in court, designing and building our house? Machines? Very doubtful!

Given these seemingly insurmountable obstacles, the prospects for a work-free society are infinitesimally small. But if we use the work-free society as an idealized limit, we can gain some insight into the likely societal changes that could come about with increased productivity and partial automation, with more of our products and services produced by machines. So, let's explore what people would do in such a strange setting.

In a work-free society, people would no longer be obliged to work for a living. If they elect to pursue it, they would have opted for desired leisure—which is not the same thing as unemployment. What might they do if they have more leisure time? More of what we do now. There are generally two directions people pursue in their leisure time: their own selves and the rest of the world.

The work-free people might pursue greater pleasure by reading and writing, experiencing and generating art, playing games, traveling. They might go after the loftier pursuits of self-actualization— learning for the pleasure of doing so, exercising their bodies for greater health, and nurturing their minds for greater personal satisfaction. On the other hand, they may simply vegetate in front of a media dispenser or take mind-expanding drugs or engage in self-destruction rooted in unhappiness.

If our leisure-laden people turn toward the rest of the world, they are likely to commune with nature, visit more with family and friends, establish and maintain relationships, and help others.

We're all aware that retirees and very rich people who no longer work follow all of these paths. But we are also aware that we, the people of the industrial nations, have collectively opted not to use the increased leisure time made possible by the Industrial Revolution for leisure. Studies show that leisure time has actually decreased. And what have we replaced these hours with? More work! Although our productivity is up and our basic needs are met, we have chosen to work even more so that we can further improve our way of life and buy luxuries that far exceed those of the preindustrial era. If human nature has its way, the human drive to own more goods and use more services long after our basic needs are met will undoubtedly continue. And therefore, so will work, demolishing in yet another way the prospects for a work-free society.

Let's now apply these observations to the Information Marketplace. People will strive to work and there will still be plenty of human work to do, but there will also be increased automation and automatization, and for many people increased leisure time.

As we have seen, we do not yet have enough information to predict whether the rise of electronic bulldozers will alter the employment rate across the economy. But to the extent that people lose jobs that become increasingly automated, there will be unpleasant social consequences that should not be swept under the rug. Jobs have been dislocated each time we have climbed the productivity ladder. We should keep our eyes open and be prepared for it to happen again in the Information Marketplace society.

The other changes are more encouraging. The increased leisure of retirees and wealthy individuals has hardly resulted in sweeping excursions to the bad. On the contrary, there is much good that flows out of the pursuits of these people. We can therefore expect a corresponding amount of good to come from a society infused with increased leisure time. Wealthier people are likely to transfer some of their wealth to take care of the less wealthy, and people with more time are likely to use it to help the unfortunate. Just the same, we must be wary of greater self-indulgence and increased mental atrophy. That affliction is sure to follow as the Information Marketplace offloads brain work, much the way physical atrophy followed the Industrial Revolution's offloading of muscle work. No doubt, we will consciously go after mental exercise to keep our brains healthy and taut just as we have pursued physical exercise to keep our bodies fit.

Looking at future economic systems from the extreme vantage point of a work-free society leads us to some confusion. Everyone will be a capitalist because all people will own machines and other income-producing capital assets. So surely they will be in favor of capitalism and will view themselves as capitalists. On the other hand, with all that leisure time on their hands, people are likely to help others on a grand scale—which would make them good socialists. Regardless of what we might call such a split orientation, we can hope that when we back away from the extreme vision of a work-free society, some of these altruistic traits will survive in the more realistic world of the Information Marketplace.

13
Electronic Proximity

A Thousand Times Closer

The Information Marketplace with its electronic bulldozers will bring unprecedented economic leverage to people, organizations, and countries. Now we assess the even greater consequences of how the Information Marketplace might affect our humanity—each of us as individuals and all people in their relationships with one another. We begin our discussion of this important question by looking at the second new force arising from the Information Marketplace: "electronic proximity."

During the Industrial Age people's physical mobility expanded tremendously, widening a person's universe of potential relationships from a few hundred village neighbors to hundreds of thousands of people within driving range. As a result, our proximity to people whom we could reach grew a thousandfold. Incredibly, the Information Marketplace will increase this range by yet another thousandfold, to hundreds of millions of people who will be within electronic reach. That is the essence of the gigantic new force we call *electronic proximity*. Because distance in the Information Marketplace is not measured in kilometers but in keystrokes and other electronic gestures, the whole scene will resemble a billion people and machines all squeezed into one electronic city block—a gigantic info-metropolis full of millions of desirable info-friends . . . and info-predators.

When greater mobility was combined with machines that offloaded physical work in the Industrial Age, a new social "middle class" of factory workers, managers, and service providers arose. In the same way, the combination of electronic proximity with electronic bulldozers will lead to a new class, which indeed is already emerging around us: *information workers*. Some people believe this will polarize society into knowledge "haves" and "have-nots"—a different distinction from the rich-poor gap we discussed in chapter II. But in nations like the United States that are steeped in capitalism this fear is baseless, because class distinctions are based on wealth, not knowledge; plumbing contractors, academics, and actors use the same country clubs. In these countries the information haves are likely to become wealthier than the information have-nots as a result of the value of their greater knowledge. That will accentuate the familiar distinctions between rich and poor rather than introduce a new social class. However, in some older countries such as India, Japan, even France, ancient cultural traditions persist in creating a clear class distinction between knowledge workers and physical workers. The rise of information workers in these places will probably accentuate this division as well as the rich-poor distinction.

Electronic proximity could have other negative effects. In the Industrial Age, changes in wealth and work patterns due to increased proximity weakened the nuclear family, while compassion and concern for community gradually gave way to materialism, a greater focus on the self, and individual goals. Another thousandfold increase in proximity will probably accentuate these trends—not a pleasant prospect but one we cannot and should not ignore.

Of course, there is just as much opportunity for good as people are brought closer to one another in their personal and professional relationships. We will have to prepare ourselves for an expansion of the positive and negative effects of much greater proximity. The Information Marketplace will affect the ways we interact with one another, the class structure in our society, the tribalism of our culture, crime, international cooperation, the reach of government, and the meaning of nations. Left unchecked, some of the changes could prove hurtful. But if we are smart about managing them, they can really help us.

Humans and Machines

Let's assess the effects of electronic proximity by using our familiar categories of interactions in the Information Marketplace: machines with machines, people with machines, and people with people.

The interaction of machines with machines is the basis for increasing human productivity. While it is the electronic bulldozers that will offload information work, it is electronic proximity that will put the many machines in contact with one another so they can work together. The combined effect of these two forces is similar to what happens on a construction site, where the real leverage occurs once bulldozers, cranes, and other machines are brought *near* one another so they can work together toward a shared goal.

The people-with-machines interaction involves things like searching for information, network surfing, recreation, electronic commerce and banking, and ordering goods. It also makes it possible for people to do everything they do now with their standalone computers on distant machines, which electronic proximity will bring near to each of us. With equal ease an economist in his Chicago apartment will be able to reach the machines at his office across town, at the Bureau of Labor Statistics in Washington, D.C., the Organization for Economic Cooperation and Development in Paris, or the Ministry for International Trade and Industry in Tokyo. Much of the success of the World Wide Web derives from this capability, as people reach easily for text, images, and programs that reside on millions of machines.

Many of the changes to our daily lives and organizations that we discussed in the second part of the book also involve interactions between people and machines, whether it's shopping, accessing our Guardian Angel, learning from a simulator, or visiting a museum. These are sweeping changes. And they will surely affect the way we will see ourselves—positively if we can enjoy the leverage that they afford, and negatively if we cannot use them.

Either way, however, as the Information Marketplace provides richer and faster brainpower services, people may begin to rely more on obedient technical servants and less on their fellow humans. Just think of how many banking customers already prefer ATM machines over human tellers. Extend that to the many services that will be offered by machines. Some additional shift of reliance from humans to machines will happen.

The dynamic of the people-with-people interaction is a bit different. Here machines mediate the exchanges between people. Telework and groupwork belong in this category. So does much of future health care, recreation, and commerce. People-to-people interaction is where electronic proximity will affect society the most, simply because it will affect on a wide scale the social and professional interactions among people. Telework alone is enough to radically alter the purchase and sale of office work across the globe.

There's no precedent for this kind of interaction in the Industrial Age, and therefore we have no guide to the issues it might raise. Saving muscle work did not directly affect the interactions among people. Although mechanized transportation helped workers get around, and the telephone revolutionized social and business encounters, people still interacted in the same way: face-to-face, or by speaking and listening.

Yet the patterns of change are emerging. As recently as ten years ago, virtually every interaction between two people in different locations took place over the phone or through the mail. Today we fax all sorts of ideas, send by e-mail complex documents without ever speaking, and negotiate with each other via a series of messages on answering machines. We also engage increasingly in online discussions, in both social and professional settings. These are some early indicators of how electronic proximity is beginning to alter people-to-people interactions. There is no telling how far these patterns will go. And when they have permeated a large part of the population, they will be felt through their broader social repercussions. Being more aware of what others are doing, being able to collaborate with fellow workers, and bridging time and space should expand the ways we feel involved and needed—friendships, student-teacher pairings, romantic linkages, and many more kinds of relationships will have greater possibilities than before. But having machines mediate human interactions will also interject a degree of isolation among people, like that felt by top executives and royalty who communicate with almost everyone through intermediaries and servants.

There is little evidence yet about the ultimate effectiveness of telework and groupwork—the areas of interaction harboring the biggest potential impact because they will affect hundreds of millions of people worldwide who do office work. Still, we can engage in some reasoned speculation about the immense change that may be upon us within a generation. What might happen if, say, 20 percent of the globe's population can play and work with one another unhindered by distance and time zone?

Rise of the Urban Villager

We expect that telework and groupwork will raise productivity further. They will also save us commuting time, human energy, and fossil fuels as we move information instead of bodies. We will still leave our homes to meet in person, but less often. Through these changes, electronic proximity will create a new and peculiar split: people of the twenty-first century will find themselves leading a somewhat schizophrenic life characterized by virtual urbanity and physical parochialism.

This new breed of people will interact virtually and in sophisticated ways with their co-workers, clients, and suppliers in New York, Tokyo, Frankfurt, and many other cosmopolitan centers. Yet, because they will be working from their homes, they will interact physically with the same few people in their families and the local grocery store, school, town hall, and recreation center for their simple yet essential needs. Meet the *urban villagers* of the Information Age—half New Yorker, half farmer—modern centaurs indeed!

Of course, not all virtual sophisticates will live in small towns, and not all small-town dwellers will operate in the virtual megaworld. City dwellers who work in corporate offices and townspeople who work locally will be around in sizable numbers. But the urban villagers will rise in the middle, and their lives will be where the grand behavioral battles brought on by the Information Marketplace will be fought. These battles are important because they represent in the extreme what all of us will face to some degree as we use the Information Marketplace to roam the world and spend more time at home.

If urbaneness dominates in the urban villagers, then electronic proximity is not likely to increase compassion, family cohesiveness, and concerns for community, because most people would agree that the physical proximity of urban living has dulled these qualities. If virtual urbaneness dominates, the twenty-first-century urban villagers may become even more indifferent to their fellow humans, pursuing the self ahead of all else even more. If, on the other hand, the villager wins the behavioral battle, then we may see a remarkable return to the values of family and friendship nurtured by the close physical proximity of adults to their children, relatives, and friends that the Information Marketplace makes possible.

At this stage, we have no clue which way the pendulum will swing. But we find it surprising and exciting that a technology that celebrates virtually bridging large distances may also have the capacity to bring people *physically* closer to one another.

Nations or Networks

The urban-villager forces of the twenty-first century will reorder the way in which people aggregate into tribes and society at large. What might happen, then, to the aggregates known as nations?

Traditionally, nations have been held together by a land mass that has economic value, a language that helps the indigenous people communicate, and a culture, history, and often religion that stem from their shared past. These forces are all losing their physical locality. Japan has proven that the economic value of land can be successfully replaced by the economic value of knowledge—which, being largely information, is brokered by the Information Marketplace. Language, culture, history, and religion are also disengaging from geographic bounds, as many more people emigrate or work abroad; one-fifth of all Canadians and one-half of all Greeks live away from their national land mass. These trends will accelerate as more people and companies see their future in a single global economy.

But even as we scatter, the Information Marketplace can help us nurture our ethnic heritage, further reducing the need for a traditional, physically local nation. As a Greek in Boston, if I had the full resources of tomorrow's Information Marketplace, I could attend cathedral services in Athens from my living room; sip ouzo and eat olives on my porch while singing native songs with my old classmates, who would also be sipping ouzo in Plaka; visit the Knossos Museum in Crete; attend an ancient Greek play at the foot of the Acropolis; shop for treasures in the Athens flea market; and watch the sunset over the Santorini volcano. I could partake of more cultural food than I have ever absorbed through the phone, the postal service, and maybe even the infrequent vacation.

Thus it may happen, perhaps within a century, that we will no longer be talking of the Greek nation as the physical country of Greece, but as the Greek Network, linking the Hellenes around the globe! Ironically, this may be closer to what the ancient Greeks meant by *ethnos* (nation) than the current land-locked interpretation of a nation.

Outrageous? Indeed. Even as I write these words I cannot believe them! But I can find little fundamentally wrong with this assumption. Neither can several national politicians and foreign ministers I have approached who, after initially reacting with indignation, have become increasingly interested in this nonsensical vision.

A Universal Culture?

Regardless of whether nations remain bound by land or become more distributed, electronic proximity will strengthen cultural ties among them. In the few decades that television has been pervasive worldwide, it has promulgated certain cultural norms, even in nations where electronic media have been specifically restricted. The collapse of the former Soviet Union may have been influenced by this factor more than we think. Television has made certain products and services universally shared, if not uniformly revered. The medium has also dramatically increased global communication; twenty-four-hour news has generally made it increasingly difficult for anyone to hide important happenings from the rest of the planet.

Similarly, the Information Marketplace will exert a blending and leveling force on the local and global cultures of the world, as people from the smallest village and largest metropolis pursue recreation, commerce, education, health, and other human activities together, whatever their nationality.

Could these new forces level out the differences among us, resulting in a universal culture that spans the globe? Not quite. These homogenizing forces can only go so far, because of the overwhelming power of ethnicity. My recurring example, the Greeks, retained their ethnicity for thousands of years, even through four hundred years of Turkish occupation. Chinatown retains its colorful rituals and deeply rooted mores regardless of whether it is in London or San Francisco. And as we see every day on the news, millions of people remain all too willing to make war in the name of their particular ethnicity. Most likely, the Information Marketplace will superimpose a thin "cultural veneer" of shared experiences on top of the individual cultures of the world. Consider how the English language has become a common bond for the member nations of the European Union—which all retain their own languages and customs.

Through this thin but universal cultural layer, the Information Marketplace will simultaneously strengthen diversity and tribalism: By getting to meet one another, via telework groupwork, and other such encounters, people will cross tribes, retaining their tribal identities while reaching out to share a universal experience. Most important, a common bond reached through electronic proximity may help stave off future flareups of ethnic hatred and national breakups by giving people another major channel for communication and cooperation, beyond trade and diplomacy.

I thought I might find some insight into this possibility at the 1995 Davos World Economic Forum, held in the Swiss Alps in February, as it is every year. Established twenty years ago through the vision and hard work of Klaus Schwab, a German professor of economics, this is a globally unique get-together. More than 1,300 CEOs representing over $3 trillion of revenue; 300 professors, intellectuals, and artists; and another 300 clergy, politicians, and prime ministers take over the small village of Davos for a week to debate everything from politics to economics and technology. Besides the intense discussions and flow of knowledge, there is amazing networking going on; side deals flow effortlessly left and right, even during the evening ball and Sunday sleigh rides. With everyone their own boss, there is no need to check with people back home as important agreements are seeded and struck. E-mail stations are in every hotel . . . and so are the security dogs and armed guards.

One afternoon I chaired a plenary session entitled "Nations or Networks" (based on the theme of the previous section in this chapter) for an audience of over a thousand people. My panelists were three powerful men: media mogul Rupert Murdoch; Michael Spindler, then CEO of Apple Computer; and Mark Woessner, CEO of Bertelsman, the German media supercompany. Not one panelist touched the controversial theme of nations becoming distributed. I was sure we had missed a great opportunity to assess the future in a gutsy way, when a written question came to the dais from the audience. I read the shaky but beautiful handwriting aloud:

When are all these new technologies finally going to let us hear from the voiceless millions of this earth?—signed, Lord Yehudi Menuhin.

No one of us could tackle the great violinist's question. It was indeed rhetorical, intended to convey what he thought the goal of the Information Marketplace should be.

I heard the sequel to Lord Menuhin's vision a few months later at the G7 conference in Brussels. The focus of this annual meeting of the world's seven wealthiest nations was to see what agreements could be reached to facilitate the orderly evolution of the Information Age across the world. Each nation sent a small delegation with the appropriate ministers. The late Commerce Secretary Ron Brown and Vice President Al Gore headed the U.S. delegation. We were met in the Great Hall by former European Union president Jacques Delors, who chaired the meeting. Strangely, we civilians were seated around the inner ring of a hundred-foot-long horseshoe-shaped arena to lead the discussion, while the politicians sat *behind* us listening!

That evening our invited keynote speaker, Thabo Mbeki, deputy president of South Africa, delivered a profound speech with dignified simplicity. He told the spellbound audience that because we are a people-centered society, the Information Age should focus on empowering people. To Mbeki, this meant that people should seize the new technologies to keep informed about the truth of their own economic, political, and cultural circumstances, and that of others throughout the world, rather than rely on the carefully crafted pronouncements of the world's governments.

Menuhin and Mbeki envision the same possibilities—that electronic proximity will enable the people in the world who "have" to see and hear firsthand the struggle of the people who "have not," empowering the less fortunate to grow out of their painful and repressed existence. Let me add that the flow need not, indeed should not, be one way, either. There is a great deal that the have-nots can bring to the haves in the Information Marketplace once a dialog gets started in earnest, such as their history, culture, experience, indigenous skills, innovative use of sparse resources, and perseverance in the face of adversity.

We can only hope that these two distinguished men hit the nail on its head. Television broadcasts have opened people's eyes to a fraction of the world's problems. The powerful and broad reach of the Information Marketplace should take us further, serving as a familiarizing force that helps people understand one another directly, through the actions of regular citizens. No longer will we be confined to the feed of governments or the interpretations of television networks.

A sterling example began in 1995 when several pacifists in the United States, Germany, and the former Yugoslavia set up Bosnews, an Internet site to help the people of war-ravaged Bosnia. Bosnians would send online messages listing what children and university students needed to keep up their education, from books to batteries. The organizers would post lists on the site, and people from around the globe would send the needed items. One American, William Hunt, a professor at Saint Lawrence University in Canton, New York, had made eight trips to Bosnia by June 1996 to deliver goods, including ten donated laptop computers to help the Bosnians expand their own computer network.

The group also set up an Internet site named Sarajevo after the national library of the former Yugoslavia in that city was razed.

People of Serbian and Croatian descent had lost a lot of their cultural record. The people running Bosnews and a few other sites put the word out for anyone with historical information to add it to the Sarajevo site. The material mounted, building an electronic repository of Serbo-Croatian culture. The electronic proximity that resulted from the site also helped relatives separated by the war find one another. The effort helped the people of Bosnia and the other five Yugoslavian republics rebuild the bridges of their past. In the process, the electronic proximity among people from different nationalities served to enhance the shared cultural layer we have been talking about through humanitarianism, compassion, and cooperation.

Spectacular activities like these are not common. Fortunately, such heroics are not essential to bring people together. The Information Marketplace, with its powerful force of electronic proximity, can help us all reach a more common understanding through widespread sharing of modest daily activities. We will visit museums together; attend plays and sporting events and street demonstrations together; chat next to virtual water coolers together; play games, bid at auctions, pursue romances, and obtain degrees together—much of it internationally.

As we do, virtual friends in virtual neighborhoods, offices, and villages will naturally share everyday human concerns and aspirations. Gradually, perhaps even unknowingly, they will be engaged in the very listening to human voices that Menuhin and Mbeki want us to do.

This vision may be wildly optimistic, but that does not mean it cannot happen. So let's seize the opportunity!

Crime and the Law

Of course, there is a downside to electronic proximity. When people get close to one another, whether in small tribes or global networks, privacy and security become increasingly important—and harder to accomplish. Today, if you are involved in a juicy controversy, television cameras will probably show up at your door and broadcast your face worldwide. The same thing will happen within the Information Marketplace. It's not uncommon for thousands of Internet users to *flame* at a person who takes a controversial position or issues a remark perceived to be offensive, dumping so much electronic junk mail onto the person's computer that he or she can no longer use it. These consequences of greater exposure will continue, resulting in increased visibility for those who stand out through their information transactions.

Info-crimes can take as many forms as there are ways to buy, sell, and exchange information in the Information Marketplace. At their most benign, they are indistinguishable from humorous hacks and pranks, where there are chuckles and maybe mild embarrassment but no victims. Beyond that level there are viruses, Trojan horses, worms, and their variants, all programs that try to sneak into your computer on a disk or over a network. Once inside, they do their ugly deed—make your machine behave strangely, steal your password and send it somewhere else, "explode" at some specified date and time by printing a message on your screen or, if they are malicious, by destroying your files. Fortunately there are ways to catch these invaders with other programs, but the possibility is always there that one of them will pass through the defenses and cause havoc.

When these program penetrations are augmented with malicious human actions, things can get worse. Any network user, no matter how modest or remotely located, could become a candidate for vandalism and defamation. In February 1996 a Canadian man electronically broke into a small provider of Internet services in a rural Massachusetts town. The invader deleted every bit of data from the firm's computers. Worse, he sent anti-Semitic electronic mail worldwide bearing the name and physical address of the firm's owner, depicting him as a white supremacist.

Electronic proximity might also expose us to theft, snooping, impersonation, the luring of minors, and threats. I'll never forget the bomb threat that came into our lab one day over one of our network connections. It said that on a certain date and time a bomb would explode on our ninth floor. It so happened that the ninth floor, all 1,500 square meters of it, was full of equipment, cables, and boxes and had a raised floor intended to hide cabling that, ominously, could have been used to hide lots of other things as well. Searching it completely was not feasible. The threat turned out to be a hoax, but it caused great commotion and shut down a building of a thousand people for a day. Fortunately, with the help of the FBI we were able to find the culprit—a teenager in San Francisco.

Imagine what crimes will be attempted in the future, especially when the advanced middleware tools and human machine interfaces will be in full swing. And these crimes need not be restricted to individuals. As we discussed earlier, information warfare could pose major problems to our society with far graver consequences if it succeeds.

Are we doomed? Hardly. In chapter 4 we saw that there's ample technology for avoiding these problems and ensuring electronic privacy and authentication. But there is also ample technology for breaking into these defenses in the ageless back-and-forth excursions of measures and countermeasures.

We will also have choices as to how much security to use in the first place, as we now do with physical security. We can buttress our homes with steel and mechanisms that open doors only at prescribed times, or we can put a simple lock on the door and leave it open most of the time, as many rural dwellers still do. Extreme security protects our property, but it costs money and causes delays and other difficulties. Simple security leaves property with little protection but is cheap, fast, and convenient. There will be a similarly broad spectrum of electronic security barriers in the Information Marketplace, with the same kinds of trade-offs.

There will also be a wide variety of info-thieves, from rank amateurs to sophisticated invaders equipped with the latest code-breaking tools. The difference from the physical world is that electronic proximity brings users and abusers much closer together. A crook from another continent, or down the street, could steal money from your savings account or intercept your direct-deposit paychecks. Personal rivals or muckrakers could find out if you subscribe to *Guns & Ammo*, check whether you were ever treated for a sexually transmitted disease, or read your love letters. Competitors could access details of a bid you've made and undercut you. A jealous colleague could scoop your work on a new patent or invention.

How would we behave if everyone in our physical world were suddenly able to show up at our doorstep, and even walk in our front door? No doubt with increased care and vigilance. To expect that human beings will behave in an exemplary fashion in the Information Marketplace is to expect that human nature will change. This belief is espoused by some techies who as young people have experienced fairness and egalitarianism in tightly knit research communities of their peers and would like to repeat the honeymoon in an idealized world of information where there are no restrictive laws.

That, unfortunately, is pure naïveté. The Information Marketplace will not be the magic path to enlightenment. It will be just another manifestation of ancient humans expending their ancient human lives in search of ancient human goals through new human tools and artifacts. Social institutions of enforcement and justice will still be needed.

To cope with info-crimes, the police and other authorities will need to adapt their techniques to the characteristics and tricks of the Information Marketplace, as they have begun to do. But the broad framework in which they perform these jobs can remain the same.

We broaden these observations to an important conclusion: *There is no fundamental facet of the Information Marketplace that calls for major changes in our current laws and regulations, including government rights and individual rights.* The main reason is the immutability of human nature. And since the angels and the devils are within us, not the technology, we can also expect the balance between good and evil forces to stay roughly the same.

No Passports, No Checkpoints

One change will be necessary, however: increased coordination of laws among different states and different nations, for the simple reason that the Information Marketplace does not recognize national boundaries.

Recall from our discussion of forbidden pleasures that in 1994 a California couple, the Thomases, posted intimate photographs on a bulletin board that they were operating. A man in Memphis clicked his mouse and downloaded them from California, where they were legal, onto his computer in Tennessee, where they were illegal. Memphis law enforcement complained, and federal prosecutors extradicted the Thomases to Tennessee, where they were convicted and imprisoned under local law.

In 1995 officials in Mannheim, Germany, deemed certain materials on CompuServe offensive. CompuServe shut down access to the site—from the entire world, because it could not for technical reasons restrict access by the location of the user.

Some national governments are making strong social decisions based on fears of foreign cultures and their perceived obligation to control the social environment of their citizenry. Singapore, China, and Iran have declared their intent to control access to Internet material. Other more "liberal" nations have shown similar inclinations. In 1995 the U.S. Congress enacted a telecommunications bill that contained a section that specified penalties for anyone offering obscene material on the Internet. The federal government was sued on this provision and lost the battle in a U.S. Supreme Court decision that ruled such restraints unconstitutional.

All these situations developed and will surely continue to develop from a mismatch of local policies among physically distant, but electronically proximate, parts of the world. The problem is vexing because the Information Marketplace crosses national boundaries and because it isn't easy for anyone to install "customs and passport checkpoints" to contain the transport of information; there are too many phone lines, video bands, satellite links, and other wireless communications.

At a recent international conference an Asian politician indicated that his country would monitor all Internet transmissions flowing through its phone lines. He contended that they would be able to snag any offensive or criminal communications crossing his nation's borders. I asked him if he actually thought he could monitor the thousands of phone lines and satellite transmissions that would cross his boundaries. He hesitated a bit but said, "Certainly." I then asked him how his operatives would know which signals were Internet traffic and which weren't. He said that they would listen for the telltale sound of flowing bits. I countered by asking if his censors would allow religious hymns to pass through uncensored. He said that naturally they would. Then I told him how easy it would be for almost anyone to encode pornographic pictures within the sounds of religious hymns that could be subsequently decoded with the right software. He said nothing, and walked away. (If hymns become censored in your country, please do not blame me. Besides, you can always encode them and transmit them within whatever ends up as permissible material!)

The point here is that unlike the few and well-defined physical national boundaries of today, with their passports and checkpoints, tomorrow's boundaries will have to contend with millions of electronic paths and millions of ways of carrying information through each path. This situation is so novel that it represents a major qualitative shift in what boundaries can and cannot do and what will and will not pass through them.

All of these new situations point to the same issue. People with different customs within a nation or across nations will want to treat differently the information that flows over the Information

Marketplace. And they should have this privilege. So what are we to do?

The encryption technologies we have discussed can be used to establish a rating and selection system. Information—whether a movie, a picture, text, or a discussion group file—can be rated by its originators or by third parties such as parent-teacher associations, or even by the government. Parents can then "lock" their child's Internet browser so it can accept only materials with ratings they deem suitable. A French couple that objects to violent American movies can block them from their son's computer according to ratings established by France's Ministry of Culture. Such a system for the World Wide Web, called PICS, was spearheaded by MIT's Laboratory for Computer Science through the World Wide Web Consortium. It was adopted by a range of computer and telecommunications companies in March 1996.

PICS is interesting because it uses a new technology to increase social options without reversing the status quo. If your government rates Internet sites, you are welcome to use their ratings. Or you can use some other group's ratings. Or you are welcome to use none at all.

As you might expect, technology can also be bent toward central censorship. For example, Iran (or Tennessee) could declare illegal the sale or use of any Internet browser without a built-in lockout mechanism that prohibits material rated as unacceptable by the government. Anyone caught with an illegal browser could be subject to prosecution. Governments can indeed take actions to exercise control on the use of the Information Marketplace by their citizenry, in spite of much noise to the contrary. Of course, they'll have to contend with the multiplicity of paths that cross their boundaries that may, in part, defeat such attempts.

A broader question here is whether authorities will use the arrival of a new technology as a subterfuge to justify changes in human policies. Technology, being a set of tools, can be used to support liberal, conservative, or any other views. Some people would like to see a new world in which governments and municipalities have no right to wiretap or eavesdrop on citizens, even suspected criminals. Others would like to give authorities control over actions and materials they find objectionable. Both sides argue that because information technology enables a new kind of behavior, new laws are needed to control the technology. This line of argument is a deceptive ploy. It lets some people exploit new technologies as a reason to get laws approved that cater to policies they favor for other reasons, when in fact technologies can be used for implementing almost any desired human policy. No one should be induced to believe that social policy must emanate from technology, as some techies would like to think. And, as we shall see in the next chapter, no one should be encouraged to believe the converse—that technology is a mere tool that can be summoned to serve the goals of a society that is somehow isolated from technology—as some humies would like to think.

Just because we are interconnected, it doesn't mean that people have the right to attack a culture that took a thousand years to build. Politicians will, no doubt, strive to protect their cultures, as they should. So in answer to our question, What are we to do? the first thing politicians may want to do is to take into account the culture and customs of their constituencies, along with an understanding of the capabilities of technology, and fashion a sociotechnical combination of policies and technologies that preserve the cultural status quo of their nation.

The second thing politicians should do is to ensure that agreements are struck across states and nations, because that is where the difficult battles will emerge and be fought. Following the models of international trade, telephony, and air transportation, all of which deal with transgressions across international boundaries, states and nations should now enter international agreements and adopt shared regulatory policies for handling trans-border information misdeeds in the emerging global Information Marketplace. This will not be easy, but it must be done. To not do so would be to deny cultural and other differences among people and to encourage chaos when conflicts occur, as they surely will.

Such agreements will help us maintain some order in the near future as we make the transition to the new world of information. But they will be difficult to maintain over the long run for all the reasons we have discussed and because of the international pervasiveness of the Information Marketplace. So the third and most important thing politicians should do is to reexamine at both the national and international levels how they might change their national policies toward information, legitimizing it as the more global and less parochial commodity that it is. Like television, it will spread and it will not be amenable to restrictive and confining policies for long. Governments might as well get some credit for making this happen in a visionary and voluntary way.

Big Brother

Whether within or across borders, we are led to wonder whether the Information Marketplace will make it possible for governments and employers to exercise their insidious control over us. Will Big Brother actually come about?

As we have repeatedly argued, the Information Marketplace, like any other set of tools, can be used for good or bad. I can imagine an information infrastructure and an accompanying political machine designed by a dictator and used to enhance and promote that dictator's purposes. All external phone conversations would have to pass through a central telecommunications authority, where they would be regulated. Anyone caught using wireless or satellite links would be severely punished. All transactions would leave behind audit trails marking which people got together, the time they got together, and so on. And because all transportation, shopping, and other activities would rely on this infrastructure, it would indeed be possible for the dictator to have good knowledge of his people's whereabouts and activities. Controlling the people under these conditions wouldn't be difficult. Just as organizations today operate their own private internal networks—their intranets—that are firewalled against the outside world, a dictator could firewall his nation's network from the outside Information Marketplace and tailor it to his purposes.

Could that really happen? Such a development is unlikely. Already, the world is moving with giant strides in the quest for massive economic growth. The Information Marketplace is a central factor in this growth and can even be regarded as the largest potential market in the world. A nation that seeks economic growth in the global economy has no choice but to join in. However, most of the control over the machinery of the Information Marketplace will be exerted by the industrially wealthy nations, which are democratic nations as well. They will decide when to turn their machines on or off, with whom to interact and not interact, and what to do with their virtual partners—not some central agency. By its very definition, this control distributed in the hands of the bulk of the people who will use the Information Marketplace runs counter to centralized control by Big Brother. Moreover, the distributed communication among millions of people under such autonomous rule will exert a further democratizing influence on all the participants. These forces, together with the worldwide trend toward democratization, will make a dictatorial information infrastructure so embarrassing and useless that no self-respecting dictator would want it.

Could the controls be far more subtle? Could an employer, for example, use Big Brother tactics to monitor or exploit its employees? Could a rich company exploit programmers and data entry clerks working in a remote and poor country? The forces and motives will be there, for example, through adoption and abuse of monitoring and wiretapping schemes. But equally subtle and not so subtle security tools and countermeasures will emerge from individuals and unions to protect workers from such misuses. Some transgressions will no doubt happen. But by its very nature as a child of the industrially wealthy and democratic nations of the world, the Information Marketplace will act as a gigantic flywheel of egalitarian customs and habits that aspiring joiners will have to adopt if they want to play! It will be difficult to hide well and for long in such a world. The chances for Big Brother's gaining any kind of foothold in such a setting are infinitesimal.

Electronic proximity will affect our lives, nations, and cultures—but if we are smart about it, only in ways that we allow. Just as we do today and have always done, we must make informed decisions about what we will and will not stand for, what we will encourage and discourage. Human nature and human needs will lead us to use the Information Marketplace as a new medium and a new opportunity to get close to new people in new ways.

This point suggests that we look deeper into human nature. How, if at all, will the Information Marketplace affect ancient human forces such as love and hate? What does this entire movement really mean for humanity? On to our final chapter.

14 Ancient Humans

Overload

Throughout this book, we have tried to assess how the Information Marketplace might transform society. And shortly, we will consider the greatest potential transformation of all, one of truly historic proportions. But no matter how powerful and pervasive a technological force may be, it will face some immutable human traits that will always act to conserve the constancy and stability of our species. Let's briefly look at two of these traits, rooted in the ancient humans that we all are, and see what happens when they are confronted by the new forces of the Information Marketplace.

The first has to do with the biological limits of our minds and the overload we experience when they are pressed. The most direct and obvious consequence of the thousandfold increase in electronic proximity is the number of people that each of us will be able to reach. That number will soon be in the hundreds of millions. Being able to reach any one of these people within seconds doesn't mean we will do so. Our computer databases can easily remember a few million items, but we cannot. We have a limited capacity for the number of human contacts we can handle at any given time and those we can cultivate throughout our lifetime. We simply cannot cement deep relationships with more than a handful of people, and we cannot remember and interact with more than a very few thousand acquaintances and business contacts throughout our lifetimes.

In effect, nature has equipped us with a limited number of "acquaintance slots," and each of us has filled many of them with physical real-life acquaintances. All the Information Marketplace can do is let us label or relabel some of these slots as virtual acquaintances! It can't increase our capacity for human contacts.

This inherent limitation gives the lie to certain politicians' claims concerning electronic town halls where thousands, even millions, of people can debate issues. It's impossible for one person to deal interactively with even a thousand people, let alone millions, whether they're real or virtual. Of course, one can broadcast one's presence from a cozy lecture room with a few guests and take live phone calls, but that's been done on radio and television for half a century, and it doesn't constitute interaction with the rest of the viewers. Any electronic town hall that purports to involve millions of people in a debate will have to rely on intermediaries—people who will engage smaller groups and then link up with one another to exchange the summary points of their charges. But that would be reinventing representative government, which hardly needs to be reinvented. It will survive intact into the Information Age, all the hype about electronic town halls notwithstanding.

This limit on the number of people that we can get to know applies to common interest groups, aggregates of friends and acquaintances, business contacts, people who deliver services—the totality of interactions in the Information Marketplace. If we try to push against it, we will hardly remember the large number of people we do reach if we ever meet them again, much as politicians forget the owners of all those hands they've had to shake. These relationships will be more fickle, more transient, and less reliable than even casual acquaintances. The Information Marketplace will offer us many more people to contact. It will therefore also force us to be even more selective in how we structure our entourage of contacts from the hundreds of millions of possibilities.

Our ability to deal with technological complexity is no less limited than our ability to handle social complexity. Our world has taken a gigantic step toward greater complexity of our surrounding artifacts in recent years. The gadgets we use have become more complicated, as manufacturers try to outdo the competition by adding features. We, the consumers, oblige with our wish to "get the most for our money." If this complexity trend is left unchecked, we will soon drown in a sea of push buttons, programming routines, and arcane rituals, all necessary to open a can of peas or hear a CD.

What we should do instead, as we discussed in chapter 12, is to use new technologies to simplify as many tasks as we can. Think of how easy it would be to use your current TV and VCR if they were redesigned with simplicity in mind. Say you want to record a specific program. As it is now, you first have to figure out the network channel number, translate it to the cable TV channel number (which is, perversely, different from the broadcast channel number!), program in the show's date, start time, stop time, tape speed, and so on. What a waste! You should be able to simply give the machine the name of the show and let it do all the figuring, as some services have begun doing. Of course, making it easier for us makes it harder for the designers and manufacturers of these gadgets. But great wealth and reward will come to the entrepreneurs who understand this and buck the mindless explosion of features with simple, clean systems, like audio components that have no wires yet link up automatically when placed together. I dream that some day we will see a fashionable rise in the value of simplicity, with equipment manufacturers and software vendors competing to offer the products with the least complexity and greatest utility.

Some people maintain that increased complexity is an inevitable consequence of the times, and that the role of computers is to manage complexity. Nonsense. Consciously throwing complexity at people (or computers) is an admission of ignorance, laziness, or haste. If something is labeled complex, that usually means we don't understand it well enough to explain it or make it easy to use. A car is an amazingly "complex" machine, yet we need only a few simple skills to drive it. The very purpose of science is to help us understand the complex world around us through simple explanations. The purpose of technology is to make new artifacts fulfill the needs of humans, not to make their lives more complicated. Our ancient human traits will ensure that we will only tolerate so much complexity; if the technologies of the Information Marketplace become too complex, we won't use them often and may just ignore them altogether.

Let's visit one other limit: the speed at which humans can handle information. The increasing speed and efficiency with which electronic bulldozers will move information in the Information Marketplace could easily overload us. Though a car is simple to use, it does put us in situations that stretch our abilities; at highway speeds at which we cannot react fast enough to avoid a deer jumping across the road. Our reflexes aren't good enough, because they evolved over thousands of years during which we could barely move at one-tenth this speed. Similarly, even though the Information Marketplace will help us handle more information more rapidly than before, sooner or later we'll come up against these limits.

More contacts than we can track, more complexity than we can handle, more speed than we can keep up with—if we let these pressures overload us, we will suffer stress and be ineffective. But we need not succumb to them any more than we would today to the potential pressures exerted by the world's 700 million telephones. In principle we can reach all these phones, but we confine our "radius of action" to the few that make sense for us. If we were to call people continually, we would feel the stress from overload. It's the same with the Information Marketplace. We need to understand our human limits and the pressures on them so that we can adjust our behavior to stay prudently within them. We will then discover that the limits do not impede our freedom and flexibility. We will still be able to pursue our pleasures, our creative impulses, our endeavors to learn and stay healthy, our work, and all the activities we've talked about throughout this book. The limits of the human body need not be limits to the human spirit.

Virtual Laughter and Tele-Friendship

The Information Marketplace will bring us much closer to other people. And because we're social animals, we will naturally try to establish relationships with them. Will such relationships have the same quality as those formed in person?

People seem to fixate on the potential of the Information Age for tearing the social fabric. Whenever I give a talk, some humie will invariably hit me with a barrage about loneliness. This happens so often that I can repeat from memory the scenario they project. Here it goes:

> An unkempt young man sits alone in front of his workstation late into the night. He plays games and surfs expertly over the Net, randomly seizing on interesting stuff. His bloodshot eyes gaze through thick glasses at a screen that constitutes his entire universe. Two pale hands on weak arms stick out of his shoulderless and sedentary frame, expertly striking at the keyboard. He has no friends. He does not know how to make love, let alone how to love or be loved. It seems that machines have replaced all human relationships in his life. What's worse, he likes it. He is totally dehumanized. He is the twenty-first-century TechnoMan.

This assault always ends with the same question: "So what are you techies going to do to prevent this terrible development?" I reflexively respond with my own scenario:

A frail old lady sits alone in her apartment. A widow for many years, she misses her middle-aged children. Their pictures are all over her apartment. She understands that they are busy with their lives and families and is reluctant to become a burden by asking them to visit her more often. She has little besides the comfort of her memories. Her few surviving friends are unable, as she is, to move around without help. She has every reason in the world to feel forgotten and dehumanized. But she does not. Tonight, as she does every night, she reaches over to her workstation, speaks a simple command, and is suddenly linked with her favorite seniors chat group. And though her eyesight is weak, she can still make out the smile here, the frown there, of her newfound friends, whose faces appear, live, on her screen as they talk about the past and share the events of their day. Except for Joe, who is cranky beyond description, she treasures these people who fill her life—all twenty-five of them. She also looks forward to the Big Event, again being discussed tonight, when eight of them will get together at a real place for a real-life party.

Human relationships will neither vanish nor be magically augmented by the Information Marketplace. For some people, like the widow or the suicidal kid (who seeks help and gets it from his virtual support group at 3:00 A.M.), the quality of their relationships and their lives will improve. Some will feel assaulted by the many info-visitors who will invade their lives. And others will bemoan the ephemeral nature of these new relationships.

The Information Marketplace at once cares and cares not for all that. Like a village market, it simply provides new opportunities for making friends . . . and enemies; for experiencing the rush of romance . . . and sex; for lending a helpful hand . . . and pushing someone off a virtual cliff; for showing one's true self . . . and hiding behind anonymity. Whether and how these relationships evolve is up to us. And because we have not yet changed in any significant way, we are likely to behave over this new medium as we have done for thousands of years, mixing our noble quests with our less noble actions. The new technologies will just help us devise entirely modern ways for pursuing these ancient goals.

Will these ways enhance or impede our relationships? More to the point, which qualities of human relationships will pass well through tomorrow's information infrastructures and which ones will not? As we have seen, the straightforward stuff of office work (text, graphs, memos, procedures) is likely to pass well through the pipes and tools of the Information Marketplace. How about human emotions, which are so central to our relationships? There is ample evidence that our emotions, too, will pass through the Information Marketplace, albeit less well.

Surely, you have laughed, become frightened, and even cried in front of a particularly vivid movie. Human history is full of relationships that flourished by correspondence. These observations, along with the friendships and romances that have appeared on the Internet, confirm that the Information Marketplace will not automatically block human emotions and relationships, as some humies want to believe. But it won't match the benefits of physical proximity in human emotions and relationships either, as some techies want to believe. We know that if the long-distance relationships of pen pals are not nourished by some form of physical closeness, they will wither. The same will happen on the Information Marketplace.

Though some unimportant business relationships and casual social relationships will be established and maintained on a purely virtual basis, physical proximity will be needed to cement and reinforce the more important professional and social encounters. Someone you know well and have already built trust toward in the physical world you have no difficulty entrusting by phone, even reaching some momentous decisions together through such a virtual connection. But you would never fully trust a new acquaintance based on phone calls alone. The Information Marketplace will "pass" human relationships only partially.

The Forces of the Cave

We have seen what passes easily and not so easily through the Information Marketplace. Let's now turn our attention to the human qualities that do not pass through at all.

Most of us think we're unique and in control of our behavior. Yet we carry the features and mannerisms of our ancestors as well as our common reflexes and human patterns acquired through evolution. The fear, love, anger, greed, and sadness that we feel today are rooted in the caves that we inhabited thousands of years ago. It was in that ancient setting that the predator's growl and the enemy's attack defined primal fear. It was there, too, that our other primal feelings became reinforced—protecting our children, enjoying the pleasure of physical contact with our mate, relying on our fellow tribespeople, and so on.

These are the forces of the cave.

And they haven't left us.

I often remember this as graduation nears and a promising student I know well is preparing to fly off into the world. I nonchalantly reach out, grab his collarbone, and squeeze it to the flinch point. I then smile broadly and say to him in a somewhat loud, menacing tone, "I am very proud of you, John, and I know that you will do extremely well when you get out of here." My spoken message is mild enough, and expected. I most certainly could have e-mailed it to him through the Information Marketplace. But my overall behavior is calculated to invoke the forces of the cave, to convey to him my nurturing and caring feelings for him and his future career together with the primal fear of unspeakable "punishments" I might resort to if he ignores my admonition! That's kind of hard to e-mail.

Could the same effect be transmitted over a very high fidelity audio and video link? No. John will not feel the squeeze. The exchange will pass at most some of the emotions that mark this interaction—not the primal forces. Okay, give John goggles, haptic gloves and bodysuit, and all the bells and whistles we can muster so that when I squeeze my glove at the remote site, he will feel my hand clutching his collarbone, he will see me, and he will hear all that I have to say to him. That won't do either. John will know that this is not the real me, but a simulation. He will probably experience a stronger emotional reaction than he would with a plain video link, but he won't feel the same emotion he'd feel if I squeezed his shoulder in person. He knows intellectually, but, more important, instinctively, that he can turn off the machine and avoid my infliction. The fear he feels is not primal. He feels a strong emotion, no doubt, but not a force from the cave!

I was the victim, actually the beneficiary, of precisely such a primal encounter when I turned thirteen. That's why I engage in such tactics today. One day my math teacher at Athens College, my Greek high school, reached over and started pulling on my suspenders. He kept pulling as he admonished me to go home and study algebra because he knew that I could do a lot better than I had been doing. At the end of the instruction, he let the suspenders snap and the primal fear that was building up within me was translated to a very real physical pain. If we were in the States, my parents would have taken him to court for abusing me. Instead, I went home trembling and opened my algebra book with an attention I had never applied to any book before. I was surprised to realize that the stuff was not all that difficult. The next day, after years of getting Cs and Ds in math, I got my first A and opened my eyes to the fun and excitement of technology. I owe my career to that teacher and the forces of the cave.

Imagine receiving a threat via the Information Marketplace: "Unless you comply with our demands, we will kill your child." Surely it would induce a great deal of fear. But it would be a rationally based emotion concerning what *might* happen indirectly as a result of the threat, not the primal force experienced in the cave when the predatory animal or enemy was in there with you. Virtual encounters, even with all the technology in the world, cannot make up this ultimate difference between the physical and the informational worlds, if for no other reason because we, the primary actors, will feel and know that we are outside the influence of the forces of the cave.

Though human emotions can pass through the Information Marketplace, albeit partially and with varying degrees of intensity, the forces of the cave can't pass at all. These primal forces are a far more important and pervasive ingredient of human life than they appear at first blush. They are the magical forces that bind parents and children, healers and patients, close associates, siblings, spouses, lovers, good friends . . . and bitter enemies. Indeed, it is difficult for us to dream up any relationship or event of importance to us that does not involve the forces of the cave! They are vital and central to our lives. The fact that they can't flow through the Information Marketplace sets a clear threshold and a boundary to the quality and extent of human bonds that the Information Marketplace can support.

What Will Be

We've now acknowledged some fundamental ancient human forces and the ways they will affect and be affected by the Information Marketplace. And throughout the course of this book we've answered the questions we raised at the very beginning. So it is time to finally consider the greatest transformation that the Information Marketplace has to offer. To get to it, let's reconstruct the growing crescendo of key discoveries we have made, which together describe "what will be."

We began with a simple but far-reaching model of the future world of information as an Information Marketplace, where people and their computers will buy, sell, and freely exchange information and information services. Our first discovery was that this Information Marketplace can indeed be built on a technological foundation: the information infrastructure. We went on to explore the many human-machine interfaces people will use to get in and out of this new edifice, from virtual reality and fancy bodysuits to the lowly keyboard, and singled out speech interfaces as perhaps the most significant and imminent. We explored the pipes that will carry our information and the ways we will bend them to give us the speed, reliability, and security we need. We also saw how a vast array of new shared software tools will evolve on this infrastructure, shifting the attention of the entire software business from individual to interconnected computers. The arrival of this foundation is certain, but it could be delayed by a decade or more if the key players continue their wars for control and their indifference toward the shared infrastructure they all need. We saw too that there won't be just a handful of winners that will survive these wars; the terrain is vast, rich, and full of challenges for almost every supplier and consumer of information to be a winner.

Our second major discovery was that the Information Marketplace will dramatically affect people and organizations on a wide scale. Besides its many uses in commerce, office work, and manufacturing, it will also improve health care, provide new ways to shop, enable professional and social encounters across the globe, and generally permeate the thousands of things we do in the course of our daily lives. It will help us pursue old and new pleasures, and it will encourage new art forms, which may be criticized but will move art forward, as new tools have always done. It will also improve education and training, first in specific and established ways and later through breakthroughs that are confidently awaited. Human organizations from tiny companies to entire national governments will benefit too, because so much of the work they do is information work.

Putting all these detailed uses in perspective, we came to realize that they are different faces of two major new forces: electronic bulldozers and electronic proximity. Each has broad consequences for society. The electronic bulldozers' effect is primarily economic, increasing human productivity in both our personal lives and the workplace. The rapid, widespread distribution of information in the form of info-nouns (text, photos, sounds, video) and especially info-verbs (human and machine work on information) is one simple way in which productivity will increase. Automatization is the other powerful effector; machine-to-machine exchanges will off-load human brain work the way machines of the Industrial Revolution offloaded muscle work. We concluded, however, that to enjoy the productivity benefits we will have to avoid and correct certain technological and human pitfalls.

To better understand the economic impact of the Information Marketplace, we explored the value of information and its consequences. This led us to a few troublesome discoveries: the huge amount of info-junk we'll have to work hard to avoid and the gap between rich and poor nations (and people) that will increase if we do nothing to stop it. Other economic consequences were less clear, like the unemployment rate over the long run, which we can't forecast even though we can foresee many new types of jobs.

Another important discovery from these explorations was the power of the Information Marketplace to customize information and information work to different human and organizational needs. To leverage this power, we'll need to make our machines considerably easier to use than they are today. With increased productivity and customization, we can look forward to a larger array of better, cheaper, more customized products and services that will reach us even faster than before. More important, by making machines easier to use and giving ourselves the ability to fashion software painlessly and rapidly, we can fulfill the promise of the Information Age to tailor the new technologies to our individual human and organizational purposes, rather than the other way around.

The second of the two major forces—electronic proximity—will increase by a thousand times the number of people we can easily reach and will bring people together across space and time. Many social consequences, good and bad, will arise as this new proximity distributes powers of control from central authorities to the many hands of the world's people. Groupwork and telework will further help improve human productivity. Democracy will spread, as will people's knowledge of one another's beliefs, wishes, and problems. The voiceless millions of the world will come to be heard and be better understood, provided that the wealthy nations help the less wealthy ones enter the Information Club. Ethnic groups may become more cohesive, as people belonging to a certain tribe use the Information Marketplace to bind themselves together regardless of where they may be. At the same time, the Information Marketplace will help shared cultures grow in nations that thrive on diversity. And though we need not change our legal framework in any major way to accommodate the Information Marketplace, different nations will need to cooperate on shared conventions for security, billing, handling violations, and other transnational issues that will surely arise as shared information crosses international barriers. On another level, electronic proximity will foster a shared universal culture, a thin veneer on top of all of the world's individual national cultures. We hope that this ecumenical property of the Information Marketplace to enhance the co-existence of nationalistic identity and international community will help us understand one another and stay peaceful.

Our exploration then brought us squarely before human emotions and human relationships. We discovered that they will pass only partially through the Information Marketplace. Physical proximity will still be necessary to consummate these emotions and recharge the batteries that will sustain human relationship between virtual encounters. Finally, we discovered that the primitive forces of the cave that lie at the roots of our emotions and passions do not pass through the Information Marketplace; deep down, our psyches know that 1s and 0s cannot love, nurture, hurt, or kill us at a distance. Because many of our most valued actions and decisions involve these forces like trust, love, and fear—the information world will not be a substitute for the physical world.

Given all these possibilities for change, we considered what might happen when they bump up against the ancient human beings that we are and have been for thousands of years. Predictably, we discovered that we will have difficulty coping with the increased social and technological complexity and overload brought forth by the Information Marketplace. Though we will be potentially close to hundreds of millions of people, we will be able to deal with only a very few of them at any given time. Yet we saw that we might be able to reduce some of these complexity problems by making the artifacts of the Information Age easier to use—a primary goal for the technologists of the twenty-first century.

The Information Marketplace will make of us urban villagers—half urban sophisticate, roaming the virtual globe, and half villager, spending more time at home and tending to family, friends, and the routines of the neighborhood. If our psyches tilt toward the crowded urban info-city, we will become more jaded, more oriented toward the self, and more indifferent, fickle, and casual in our relationships with others, as well as less tightly connected to our families and friends. If we tilt toward the village, we may be surprised by a resurgence of more closely knit families rooted in our tighter human bonds. Indeed, if we use it correctly, the Information Marketplace can be a powerful magnifying lens that can amplify goodness—employing disabled and home-bound workers, matching help needed with help offered via the Virtual Compassion Corps, and helping people learn and stay healthy, among many other possibilities.

Reflecting on our exploration, we also discovered that people will exploit the newness, vagueness, and breadth of the Information Marketplace to support their wishes and predilections, whatever they may be. Some proclaim that the world of information can stand out only by offering educationally and culturally rich opportunities that will benefit humanity. Others will use the Information Marketplace as a new battleground for the familiar disputes—capitalism versus socialism, greed versus compassion, materialism versus spiritualism, practicality versus abstraction—all suitably described as "new" issues. As in the case of money, there is hardly an event, action, or process that is not linked to and affected by information, so such arguments can sound plausible. But they should not deceive us; the discerning eye will distinguish that which is likely from that which is merely possible.

The wise eye will also see that the Information Marketplace is much more influential than its parts—the interfaces, middleware, and pipes that make up the three-story building on which we stand. Once they are integrated, they present a much greater power—the power to prevent an asthmatic from dying in a remote town in Alaska, to enable an unemployed bank loan officer to find and succeed at a new form of work, to allow a husband and wife to revel in the accomplishments of a distant daughter while also providing emotional and financial support. These powers are far greater than the ability to send an e-mail message, or to have five hundred TV channels—and their impact goes well beyond affecting us economically.

The Information Marketplace will transform our society over the next century as significantly as the two industrial revolutions, establishing itself solidly and rightfully as the Third Revolution in modern human history. It is big, exciting, and awesome. We need not fear it any more or any less than people feared the other revolutions, because it carries similar promises and pitfalls. What we need to do, instead, is understand it, feel it, and embrace it so that we may use it to steer our future human course.

We could stop here, after putting all these discoveries together, satisfied and impressed with our overarching vision of a third socio-economic revolution. However, if we look even deeper at the bold and historic imperative that the Information Marketplace calls us to embrace, we will see all three revolutions as part of a far greater movement, well beyond combines, steam engines, and computers—a movement toward a new age that may liberate the total human potential within each of us.

On to our final discovery.

The Age of Unification

No matter how consumed we become with our daily pursuits, we are never more than a mental half-step away from a much greater awareness of our existence on this planet. This was starkly brought to my attention one night some years ago, and it was then, after an impromptu bull session, of all things, that I began to finally articulate in my own mind our final discovery catalyzed by the Information Marketplace.

It was 3:00 A.M. and my home phone rang. One of my dedicated young colleagues was panicked. He and a handful of students and research staff members had just discovered a virus that threatened the lab's computers. I hurried over, imagining the worst. But by the time I got there they already had it under control; the threat turned out to be less serious than anyone had expected, and we all felt relieved that the laboratory's precious files and programs were intact. Of course, at that hour, and in anticipation of an all-nighter, the pizzas had been ordered. Maybe it was the lateness of the night or the ominous invasion from outside that prompted them, but whatever the cause, they began heatedly debating the eternal questions:

> "I believe in something powerful, but not necessarily a man in a beard."
> "Religion is the opiate of the people."
> "I am an agnostic."
> "I believe in God but not in these priests, who are pretty bad salesmen anyway."
> "I am an atheist, and I am not ashamed to admit it."

"If God is just, how come there is so much suffering in the world?"

"Maybe God is a computer."

"No, God is a committee."

"Miracles? Give me a break!"

And on and on.

I was about to leave when one of them said, "Hey, let's ask Michael what he believes in." An uncomfortable silence descended as I groped for an excuse to avoid the subject. But there was no graceful way out, and besides I have yet to meet the professor who can resist a lecture . . . on any topic. So I let it rip:

"You're debating the classic conflict between faith and reason. Many of the world's best minds and countless people have spent their lives trying to prove or disprove the existence of God. In other words, they tried to use reason as a higher force to justify faith. Similarly, many theologians and even some scientists have tried to use religion to explain nature, science, and why we are here. That's trying to use faith as a higher force to justify reason. To me, neither faith nor reason can be subordinate to the other. They are like the engine and the wheels of a car. You'd better have both if you want to get anywhere.

"Like it or not, reason is full of inconsistencies and holes. Russell's Paradox is the perfect example." One very thin and frizzy-haired eighteen-year-old, who rose three inches above my six-foot-four-inch frame, volunteered that he didn't know the paradox, so I repeated the classic quandary, first posed half a century ago by the famous English philosopher and logician Bertrand Russell: "'Imagine a village where the village barber shaves exactly all the villagers who do not shave themselves. Does he shave himself?'

"It is illogical to answer yes because then he would be violating the rule by shaving someone who shaves himself. And it's just as illogical to say no, because then he would be violating the rule by disregarding someone who does not shave himself.

"So this perfectly logical and reasonable rule ends up in inconsistent nonsense. How do you know that the next logical rule or argument, perhaps one you use to support or debunk God's existence, isn't full of similar holes and inconsistencies?"

One of them burped pointedly, but most seemed to be absorbed by the argument. So I continued: "Faith is full of holes, too. If you jump off the terrace of this ten-story building in the belief that God will take care of you . . . well, that would be a gigantic leap of faith!"

I paused for chuckles. There were none.

"Questioning our faith with reasoned argument is just as un-natural as defying logic with blind faith. But if we embrace both faith and reason and let them work together, exploiting their strengths and avoiding their weaknesses, then we have something that is more powerful than either. Look at our bodies. They have been designed to do just that. The cerebrum is full of neurons that process information; it is the very picture of logic at work, with every reflex and every idea having a cause and a result. No doubt it is the basis for our being so impressed by logical thinking. But then there are glands and secretions that trigger our emotions, our pas-sions, our fears, and our beliefs. These two are at play all the time. Use one and ignore the other, and you are no longer human. I'm sure there have been times in your lives when you have wanted to take an important step, like enter into a serious relationship or buy a house, and your reason objected. So what did you do? You recal-culated the consequences and bent your logic until it agreed with your passions."

I wasn't sure I had impressed them strongly enough. I no longer held back.

"So it seems your impeccable logic, which you consider your most powerful asset for your careers as techies, works part-time as a whore to your passions! And how many times have you done the opposite—suspended your quickly summoned intuitions about someone you just met or a place you just visited, when confronted by the irrefutable reason of your own observations?"

They winced.

"I see your disbelief. You really think that reason is supreme. You trust it. You use it. You are proud of it. I sincerely hope that none of you comes face-to-face with any serious misfortunes, like a terminal illness or the death of someone really close. When people meet with such tragedies, they get little comfort from reason. They desperately need another force that will sustain them and give them strength. That force is faith.

"The lesson here is that we should not single out faith or reason as superior. Let's instead accept that we need both and that we are better off if we learn to use them in concert, thereby strengthening ourselves with their combined power.

"Which is a long-winded way to answer your question: Yes, I do believe in God. And I consider it a waste to use reason to question God's existence. Instead, I try to use both faith and reason to tackle the good and the bad that life brings my way, sometimes failing and sometimes succeeding, like all humans."

The freshman from California tossed her hair and said, "Cool!" The burper asked loudly, "Did you use faith or reason on us just now?" Another student grimaced and mumbled that I sounded like his father. And my frizzy-haired protégé said, "I know what you're getting at. You are on your favorite hobbyhorse about integrating the techies and the humies, cleverly disguising it as an argument about integrating faith and reason."

I admitted that they were way too bright for me. I thanked them for protecting the empire from the evil virus and bid them good night.

As I drove home I mused that my frizzy-haired friend was right. I have always been interested in pursuing the maximum breadth of knowledge that I could achieve, while striving to maintain sufficient depth to avoid dilettantism. And I have always been intrigued by bringing opposites together: Faith and reason. Art and technology. Creativity and analysis. Humor and seriousness. All apparent contradictions begging to be treated in isolation yet harboring in their union a power greater than the power of each part. So it is with the contrast between humanists and technologists.

The humie-techie split is a fairly recent distinction. It started stirring during the Renaissance but blossomed during the Enlightenment—that eighteenth-century movement that sought the interrelationship among faith, reason, nature, and man. Before the Enlightenment, people looked at these four pieces as an interconnected whole—some would say as a confused whole, where challenging one, especially faith, with any of the others, especially reason, could easily cost the challenger's life. The "enlightenment" of the people of that period helped them separate science from religion, morality, and the literature of the ancients. People could then pursue science independently, letting it drive them where it would. This dissociation of reason came, as if by plan, just in time for the Industrial Revolution. Or perhaps more accurately, it caused the Industrial Revolution, by letting science flourish and become translated into the technological innovations that farmed the land, ran the factories, and transported the people and their goods.

However it happened, the increased wealth made possible by the Industrial Revolution reinforced the "correctness" of the split.

The principal consequence, materialism, became a new god. So did man, with his increasing preoccupation with self. Reason, the original separatist, became yet another god with its own principal agents, the scientists and technologists. As these practitioners became increasingly recognized for shaping the future, and as they increased their specialization, they moved further and further from the humanists, polarizing the split of the Enlightenment into the *humie-techie split*. As I look at my own institution, MIT, from this perspective, it makes good historical sense that it was formed 150 years ago, after the end of the Enlightenment and immediately after the first Industrial Revolution—just in time, it would seem, to consciously legitimize the importance of technology to the world, while unconsciously accentuating the techie-humie split.

Today, this split is so ingrained in our society and in us that we accept it as a universal truth. It starts from our earliest school days and is even considered cute. Children who like math are expected to dislike literature, and vice versa. Parents reinforce the polarization. "Mary is like me. She hates numbers, but she is great with art." Or "Jimmy is always tinkering with toys, just like his father tinkers with electronics and cars. He'll be a great engineer." Later on the polarization continues. Young adults who go to college specialize in either the humanities or science and technology. Institutions add to the conspiracy by focusing on one side of the divide at the expense of the other: "The student who underpays at the supermarket is either from Harvard and can't count, or from MIT and can't read." Later in life the divisions become calcified. The artist scoffs at the engineer's insensitivity. The engineer laughs at the artist's sensitivity. Humanists defiantly repeat inside their heads and over loudspeakers the mantra that technology is a servile art to human purpose, while techies assert with equal defiance and repetition that humans are merely meat machines.

Why make such a big deal out of such a natural split? Surely we need specialties. Why get upset over the playful bantering among them? Please hold on, and you shall see.

Most humanists still think that technology is like wood and nails. They believe that people should first decide what objectives they want to pursue, based on the best humanistic thinking they can muster, and then go out and buy the technologies needed to construct their plans. In the days of the steam engine and electricity, that was somewhat the case. However, in the age of nuclear power, synthetic drugs, and information infrastructures, this notion is no longer valid. In our increasingly complex world, technological and social issues are becoming more and more intertwined. Whether designing a twenty-first-century automobile, deciding where to locate a nuclear plant, planning the growth of a city, leading a large organization, setting the privacy policies of a new health care system, or deciding where to live, we are all increasingly confronted with many mutually interacting technological and humanistic issues.

More important, new human purposes often arise out of new technologies. How can you know that you have the option to build shelters for the poor if you are completely unaware of hammers and that they can be used to build houses faster, cheaper, and better than clay and leaves? How can you set out to match the "help needed" people with the "help offered" people on a worldwide basis if you don't know about information infrastructures and electronic proximity and how they can make such matches possible?

On the other side of the divide, scientists and technologists often become so preoccupied in their quests that the endeavors themselves become their principal goals. "Don't give me all that soft stuff about human purpose. All I want to do is pursue scientific truth in the lab. Let somebody else deal with purpose and the administrative details and all the other nonsense that keeps science and technology from advancing." Other techies offer less stereotyped arguments: "Radar was invented as an implement of war. No one could have anticipated then that forty years later it would become the cornerstone of the world's air transportation system. Therefore it is fruitless to worry about purpose." Such techie views are as one-sided as the humie views we discussed earlier. Good technological innovation arises out of human purpose just as often as good human purpose arises out of a knowledge of technology. The techie-humie split has hurt both of these avenues to progress.

The growing division between techies and humies—really among the pieces within us blown apart by the Enlightenment—goes well beyond limiting our ability to comprehend and manage the complexities that surround us. The trouble it causes is bigger than it seems, affects all of us, and can be heard increasingly from all kinds of voices. The world's people, having drifted away from their wholeness in the pre-Enlightenment age, sought comfort in the good life that the material gains of technology would bring. Having largely achieved these gains in the industrially rich world, we have discovered, often painfully, that *something* is still missing.

Young people started telling us this by turning to nature, searching for spiritual directions, and moving to drugs and other artificial pleasures as their adult role models veered toward the amassing of wealth, greater self-interest, and greater pleasures. Psychiatry flourished and moral compasses increasingly began to point in all directions. The dissonance within us got louder.

This unrest got translated to a dissatisfaction with government and many people's conviction that technology was the primary cause of all these problems in the first place. That's as ridiculous as a society of beavers concluding that the dams they construct are the cause of their unhappiness. Like beavers, people are part of nature and build things for their purposes, which are just as much a part of nature. To accuse technology of bringing ills to humanity is no different than accusing the hammer you built of smashing your thumb. Of course it did, but you wielded it. And even though it caused you some pain, it also helped build your house. The alternative we sometimes hear, of stopping technological progress to save ourselves from further trouble, is just as unnatural, for it shackles the human spirit by keeping us from exploring the unknown.

People's discontent and search for purpose is a symptom of a deeper cause. I believe that we are really longing for a way to blend those old forces that held us whole for millennia, weaving a strong net around reason, faith, nature, and man, until the Enlightenment came and yanked them apart.

The Information Marketplace, if left unchecked, will further aggravate this polarization, past what we may be willing to tolerate, and may well increase human dissatisfaction to the point where we will seek radical and wholesale change. If the physical technologies of the Industrial Revolution were responsible for setting the technologist apart from the humanist, then the information technologies with their disembodied virtuality and their disregard for physical proximity will further aggravate the split. Humies who already look with some contempt at the negative consequences for humanity of the factory and the automobile will double their contempt when confronted with the impersonal and remote processing and transporting of information, let alone the unbearable fake of virtual reality. Meanwhile, computer technologists and information specialists, who already feel sorry for the old-fashioned engineers who must remain at work within the constraints of the physical world, will barely see the humanists across the great divide.

The rest of us will feel this heightened aggravation of the techie-humie split as a further reduction of our ability to cope with the increasingly complex world around us. Greater polarization among the parts of us dissociated by the Enlightenment will widen the disparity between the mechanics of our everyday lives and our deeper sense of human purpose. We will feel increasingly oppressed by our own dissatisfaction.

We don't have to sit quietly by and observe all this. We can and should act. No doubt some people will foolishly strike at the technologies—try to break the hammer that bruised their thumb. But most of us will marshal our human energies to seek a new course. To move beyond this impasse, we will first try to understand it. We will then realize that pulling apart and isolating the various pieces of our self, as the Enlightenment caused us to do, was our historic process toward that goal. Just as the psychologist isolates the trouble spot and keeps looking at it before integrating it with the rest of the self, and just as the systems engineer isolates the faulty subsystem and analyzes it before reintegrating it with the whole system, we humans have been "studying" through our experiences the isolated pieces of our being, and feeling the consequences. But now we are confounded. Our world has become like a huge ball of intertwined red and blue string. Sooner or later, we will realize that we cannot begin to understand it by focusing only on one color. We can no longer proceed to make decisions as either techies or humies, ignoring the other strand of life and the way the two are interwoven. It will finally dawn on us that if we do continue in this way, we will pay dearly for our insularity with more of the malaise that is already developing around us. More important and more to the point, we'll miss all the good that can flow from reuniting the technologist and humanist that are within every one of us.

That is the big challenge before us at the dawn of the twenty-first century: to embark on the unification of our technology with our humanity.

That doesn't mean that everybody will need to learn calculus and Latin. Nor does it mean that we will eliminate our various specialties, for we will still need them to cope with the complexities around us. It does mean, however, concerted action from all of us toward embracing, understanding, and accepting our two halves, whether they are within us or around us. How might this be done?

First, the high priests of the split will have to provide a good example by changing their ways: Humanists will have to shed their snobbish beliefs about the servile arts, and technologists will have to shed their contempt for the irrelevance of humanistic purpose and teachings. And both sides will need to actively bridge the techie-humie gap in their reflections and in their actions. Second, parents and educators will have to help young people (and themselves) learn about and experience the exciting and practical prospects for human wholeness. Toys, childhood stories, and the examples they set can go a long way toward instilling the values of integrated thinking in young minds. Curricula in high school and especially in the university will have to change in a big way, combining techie and humie knowledge and approaches in the teaching of the arts, the sciences, humanities, and management. Consider, for example, a hard-core techie field like computer science. A class could undertake as a one-year project the task of designing a computer system for a real nursing home, with all the techie-humie issues that this would entail. Or consider a hard-core humie field like literature or history; it could be recouched to explain how a hot current problem like clashing national cultures on the Web has been "handled" in its various incarnations from time immemorial.

Ultimately, all of us can contribute through our everyday actions and through our professions. Businesspeople can create new jobs across the humie-techie divide. They can begin sending the techiest of techies to make sales calls—they may not be as smooth as trained salespeople, but their experiences will certainly cause products to be improved. Politicians can begin learning about technology and using it in their plans, because most of them come from a humanistic background. They can also help enact laws that will facilitate humie-techie synergy. All of us, regardless of specialty, can try to perceive and interact with our world in its full techie-humie splendor by reading, observing, and learning about "the other side" and by searching for opportunities to combine these extremes toward profit or self-satisfaction. In short, all of us need to recognize that we are more than we thought or were taught, and that the pursuit of our broader capabilities can hold great benefit for ourselves and for society.

Oddly, the new world of information, a serious "culprit" in this mounting polarization and dissonance, may provide some help in bringing our torn-apart selves back together. This notion is consistent with our recurring observation that technology and human purpose work best when combined in support of a common human goal. Parents and teachers can use the Information Marketplace to help young people learn about human wholeness by exploring the rich world around them, ignoring whether the Web sites they visit have techie or humie colors. We can use electronic proximity to reach across the great divide, linking techies and humies in the workplace, in our daily lives, and in our leisure time—toward more effective and better-targeted organizations, more worthy projects, like the Virtual Compassion Corps, and ultimately greater purpose and satisfaction for all the participants.

Whether inside or outside the Information Marketplace, these imperatives to reunite our humanity with our technology are not easy tasks, because they call for wholesale change in human thinking and in a behavior that has taken hold of us for several centuries. It will take great effort and perseverance to undergo these changes. But it's worth it. Unified, we shall thrust ourselves and our world forward in ways that will satisfy and pay tribute to the new wholes that we will be.

Then this new Age of Unification will rejoin within us faith, reason, nature, and humanity, paving the way for the Fourth Revolution, beyond human artifacts and their consequences, aimed inward at understanding ourselves.

Unification

Techies,
Mind your prescriptions for the world.
Humies,
Tone down your fears of techno-change.
Step outside your precious castles
Look within before the split
Fill the space that makes you whole
Enjoy the sunset
And the wheel.
Argue from logic
And emotion.
Technology is humanity's child
As is our quest for human purpose.
To love them is to love ourselves.
There are no differences.
Only labels.

Afterword
What To Do

As I traveled around the world to discuss with hundreds of different audiences the hardcover edition of *What Will Be*, I was struck by how many times I met the same handful of reactions: Many people feared that the warmth of human relations will give way to the cold efficiency of computers, that people will lose their jobs, that they'll be overwhelmed by too much complexity and too much information. Others felt helpless, as if they were facing an unstoppable steamroller. Still others looked eagerly to the future and how they could exploit the Information Marketplace to better their professional and personal lives.

Whatever their individual opinions, almost everybody voiced a similar refrain: "Don't just tell us what will be. Tell us what to do." So I decided to do just that by adding this afterword to the paperback edition.

The answers are provided in the form of six action agendas. The Agenda for Individuals in the Industrialized World is aimed at citizens of wealthy nations. The Agenda to Help the Poor suggests ways to benefit the people of developing nations and the poor people of developed nations. The Business Agenda is directed at executives who want to improve their organizations and at daring entrepreneurs who wish to start new companies. The Techie Agenda is for the computer and communications technologists who construct the software and hardware systems we all use. The Humie Agenda is for the humanists who are concerned about the social consequences caused by big technological change. Finally, the Government Agenda is for citizens and politicians worldwide who want to know what role their governments should play in the Information Marketplace.

Good luck to all.

Agenda for Individuals in the Industrialized World

Are you worried about losing your job, your privacy, or your human relationships in the oncoming wave of new technology? Don't be. None of these often touted "dangers" is a necessary outcome of the Information Marketplace. Sure, life will change. But you, a citizen of a wealthy industrial nation, can control how the change affects you. Indeed, if you prepare yourself for the coming revolution, you stand a good chance of obtaining a better living and an even more rewarding personal life. You might start with the following actions.

1. **Accept and explore the new world of information.**

The Information Revolution is already upon you. It will affect you as dramatically as the Agrarian and Industrial revolutions. Ignoring it, dreading it, or even genuflecting before it will get you nowhere. It will seem as silly in retrospect as if you had feared or worshiped the plow and the automobile. Like these earlier innovations, the Information Marketplace will enter our lives and become indispensable through the useful things it will make possible. If you have not already done so, your best strategy is to begin exploring it to help you communicate with friends and associates, pursue hobbies, shop, learn, stay healthy, do business, let your voice be heard, listen to other people's voices, and a lot more. The idea for now is to have some fun and pursue some useful activities on the Information Marketplace while you educate yourself about what is and is not possible. The sooner you do, the sooner you will be able to exploit it to your advantage while guarding against the inevitable problems it will bring. And you will build up firsthand knowledge of what is feasible, without having to rely on or be misdirected by the ever-present hype.

2. **Support the Information Marketplace.**

Just as we need an extensive and sturdy road system to get the most out of our cars, we also need a well-designed information infrastructure to take the best advantage of new technology. That's why you should demand, through the political process and with your consumer power, that the infrastructure underlying the Information Marketplace be built as effectively as possible. Be prepared to pay some taxes toward that goal. Unpopular as this prospect may be, governments exist to tend to the common good, and information infrastructures will become as essential to our national well-being as highways and telephone networks. Notwithstanding rhetoric to the contrary, it may not be possible—or desirable—to relegate the infrastructure entirely to the free market, any more than it was with the highway and phone systems when they were constructed. As we discuss further in the Government Agenda, a more careful building process will create a more useful system.

3. Extend your current skills.

In the Information Age, many individuals will be able to offer work from their home or nearby work center: clerks, teachers, doctors, nurses, secretaries, executives, engineers, graphics designers, writers, telemarketers, real estate brokers, accountants, and many others. Suddenly, they will be able to reach clients worldwide. And to do it, they won't have to become computer experts or be "empowered with greater knowledge" as the current hype suggests. Salespeople, for example, will use their microphones and screens to build customer trust and answer the age-old questions about price, discounts, reliability, acceptance policies, and so on. Their success will still depend on extolling and matching their products to the needs of human buyers. And professionals will deliver the services of their specialty, which they have labored hard to learn, as they have always done. They will all have to make some adjustments to suit the new medium, but these will not be difficult to master, compared with the principal skills.

So ask yourself if what you do or sell today could be done or sold tomorrow over the Information Marketplace. If you say yes, then ask yourself if doing so would benefit you by broadening your clientele, by serving people in other locales who may be willing to pay more, by collaborating with someone at a distance, or for any other reason. Test-drive the new developments reviewed in chapter 4, especially automatization tools, e-mail and e-forms, groupwork and telework, and hyper-organizers and finders. They should help you work more effectively. While you're at it, try the emerging uses reviewed in part II, from music services to guardian angels and automated tutors. You may see in these scenarios ways that directly help extend your skills. If you do, you should seriously consider deploying your skills over the Information Marketplace, perhaps on a part-time basis to test the waters.

4. Change your skills.

If you fear your current skills may become obsolete in the Information Age, adapt them or learn new ones. Stockbrokers who worry that automated stock transactors will displace them can become financial advisors who may even offer such automated services to their clients along with their human advice. More broadly, if the goods or services you sell can be directly and easily purchased by merely specifying them (like one hundred shares of IBM or a copy of *What Will Be*), chances are that direct online sales will steal customers away from you or outdate your current system of dealers, franchisers, brokers, or other intermediaries. In such cases, a frictionless market, in the words of Bill Gates, will prevail between sellers and buyers with a minimum of intermediation simply because that will be the least expensive way to do business.

Remember, however, that as we discussed in chapter 11, most goods and services will not fall in the category of a frictionless market, because to buy them, people will need help finding them, and will rely on intermediaries for information, advice on choices, help in transactions, and, more generally, people in whom they can place their trust. So, if you face this risk, consider becoming one of this new breed of middlemen. If you start now, you'll be one of the first people in these new professions. Just to spark your imagination, here are a few that come to mind: telework brokers, group-work conveners, info-pilots who help us navigate through info-junk, info-tailors who customize information from news to insurance policies, hyper-secretaries who create hyper-organized records of meetings, info-artists skilled in the new media, authors and teachers who create exciting and novel educational materials, virtual travel agents, marketers and salespersons for every conceivable info-good and service, electronic department store operators and employees, and much, much more.

5. Don't be fooled by technological imperatives.

A time of rapid technological change is an opportunity for all sorts of people to try to advance their agendas . . . typically under the cloak of modernity and technological inevitability. Sometimes these efforts will be purposefully malicious or fraudulent. Many times, however, they will be the inadvertent side effects of legitimate intentions. For example, telemarketers will consider it their birthright to assault you daily and to try to glean your preferences and habits. Don't automatically give in to their demands on your privacy and your time. Insist, via the political process, on redress, such as requiring telemarketers to furnish their name and a way for you to permanently block them from calling or e-mailing you again.

This can be done overnight with the introduction of a public key cryptography system and digital signatures, or the requirement that all telemarketing messages be tagged with an appropriate code so your e-mail filters can automatically discard them.

Even technological changes touted as boons to law and order should be scrutinized to make sure they don't "hide" planned or inadvertent intrusions: Requiring every citizen to register their cryptographic keys with the government may look like a continuation of current governmental practices for wiretapping criminals, but would, in effect, dramatically skew the balance on citizen and organizational information potentially accessible by the government.

6. Don't tolerate misuses of information systems.

Most misuses of technology are wrought by us, not by inanimate machines. Don't accept the claim: "I didn't do it, the computer did." For that matter, don't tolerate any of the misuses identified in chapter 12: The additive fault, where you do all that you did before, plus some new stuff to look and feel modern; the perfection fault, where you spend valuable time fussing with the appearance rather than the content of information. Also beware of software that purports to be intelligent but gets in your way (fake intelligence fault) or software that is overly complex (excessive complexity fault), loaded with excessive features (feature overload fault), or controls you rather than the other way around (machine-in-charge fault). Try to avoid these corruptions, or correct them or advise the people perpetrating them.

In the end, use your common sense. You should let no person, organization, or policy affect you in an undesirable way. And if you don't understand clearly how something might affect you, insist that it be explained in the plainest way possible. I assure you that there is no aspect of computer and communications systems, or any related procedure or policy, that cannot be explained so that you can understand it.

7. Insist on adequate privacy, authentication, and payment schemes.

The ubiquity of computers and networks, and the immediacy of electronic proximity, suggest that all sorts of new good and evil will be at your doorstep each day. Stay vigilant about privacy, authentication, security, and payment arrangements—or the lack thereof. Demand that your politicians and service providers ensure your safety, privacy, and other rights. For example, the fact that we will be interconnected more tightly through the Information Marketplace does not give license to any person or organization to invade, distribute, or sell sensitive information about you.

Part of the responsibility falls on you. Be careful. Before you divulge any information, find out about the further chaining of it—where it will be sent or sold. If the answers are not satisfactory, refuse to supply the information and choose a different venue. If your rights are violated, pursue court action if appropriate. You should also pursue your rights with groups that misuse information about you that has somehow already leaked out. Demand that they cease their activity and be prepared to prosecute if they continue their transgressions. It is up to you, the citizen of an industrialized nation, to ensure that information about you is also regarded as a first-class citizen, treated as seriously as your precious tangible physical property, not a free good. If you find that the laws about privacy are inadequate, insist that they be changed.

8. Stay actively informed.

You try to stay informed about the food you buy, the water you drink, the education you're getting—all so you and your family can safeguard against new dangers and make the most of vital resources. Add the Information Marketplace to your list. Be sure that you and your children stay informed about changing technologies, and the social changes they portend. Read books, listen to the radio, watch TV news. Ask your legislators if you should worry about any forthcoming changes in the new world of information. If they say they don't know, suggest that they find out. If they identify a problem, ask them what they plan to do about it.

9. Face your fears head on.

A growing Information Marketplace does not mean you will lose your job. And it does not mean you will be relegated to a lonely existence (a widespread misconception I hear almost everywhere I go). You will still live with physical neighbors and you will still meet friends, go out to dinner, admire sunsets, ski, sail, play with your kids, caress your mate. Nothing is changing in your familiar physical world. You will become lonely if you elect to become so. You are in control. Moreover, if you are lonely now, chances are you can reduce your loneliness by acquiring friends over the Information Marketplace.

You won't be overwhelmed by complexity and "information overload," either. Since time immemorial, humans have placed self-preservation above all else. If you sense that information is surpassing your tolerance threshold, ignore it. Toss it. That's what most people do now if they get overloaded. So, your humanity won't be automatically endangered by loneliness or excessive complexity. In fact, your humanity may even be enhanced through a set of new tools that extend your reach beyond its current limit. You might even look forward to the reunification of your inner technical and humanistic selves. Above all, remember that the forces of the cave (chapter 14) will always predominate. The primal human emotions most fundamental to your life and work cannot pass through the Information Marketplace and can only be transmitted through physical presence.

Agenda to Help the Poor

Poor people cannot afford the technology to access and use the Information Marketplace. Most of this agenda is, therefore, aimed at the rest of us, who can find ways to help the people of developing nations, and the poor people in rich nations, to board this fast-moving train. For if they do not, we can expect a widening gap between the rich and the poor, which will only increase conflicts, even bloody revolutions, between the have-nots and the haves. In discussing this topic, people often tell me that the poor would benefit more if the help provided went toward food or health care rather than to information. That's like arguing that fish is directly more useful to the poor than learning how to fish. It's only true in the short-term. If the poor can join the Information Marketplace, they will be able to contribute their inexpensive labor to the industrial economies, ascend the wealth ladder, reach a better standard of living, and create new markets for all producers. Remember, too, that the poor can teach quite a bit to the rich. So while the Information Marketplace may not be the most direct way to help the poor, it can contribute greatly to helping them out of their poverty.

1. International organizations: adopt a bold new mission.

The poor will not spontaneously become interconnected with the global information economy, because they do not have the means. To take just one example, the World Bank borrows some $15 billion a year and lends it to the developing world mostly for agriculture, roads, nutrition, education, and other structural improvements. It is now contemplating how to exploit the new technologies by assembling and distributing its vast knowledge worldwide, through a knowledge network. While that is a worthy endeavor, people can't eat knowledge. They must earn money to buy food and build their own infrastructures. If the poor can be helped to exploit the Information Marketplace to create work, then they can earn money and really begin to help themselves.

The World Bank has a great opportunity to lead other global organizations by inviting proposals from the developing world and boldly earmarking a quarter to half of its funds, for a decade, toward effectively bringing the developing world into the Information Marketplace. Connection to the Information Marketplace should be a top priority. Programs for literacy are the clear starting point: if people can't read they can't use the Information Marketplace effectively, though they may still participate by using speech. Programs that help the poor learn how to cultivate crops and improve health care are obvious next steps, as are the crucial training programs that will prepare them to use the Information Marketplace in profitable ways.

Spending $200 million a year for ten years in, say, Sri Lanka for information kiosks, terrestrial communications, training programs, software, and equipment could start a chain reaction that would eventually move a good part of that country's workforce and schoolchildren onto the Information Marketplace. The World Bank has the expertise to monitor such projects and ensure that they achieve their stated goals. Corresponding steps could be taken by other organizations with a global reach, like the United Nations, the World Trade Organization (WTO), and the G7 (now G8) group of industrially wealthy nations. Governments like the European Union, the United States, and Japan could shift some of their foreign aid to the same goals. A total contribution equal to just 2 percent of each poor nation's GNP could make a big difference.

2. Commercial organizations: donate equipment and services.
Through new and existing philanthropic programs, communications companies should donate or sell at low cost backbone and local communications bandwidth to developing countries and to less privileged towns and neighborhoods within rich countries. For example, the Low-Earth-Orbit Satellite system that will soon whip around the earth carrying commercial traffic will be essentially idle when they fly over the developing world. For very little added cost these systems can operate freely or cheaply during these intervals. Bill Gates has already indicated to me that he would consider donating such service from the Teledesic Low-Earth-Orbit Satellite system, of which he is a major stockholder, once it gets into orbit.

Similarly, geosynchronous birds such as the ones planned by Hughes and AT&T to serve industrially wealthy regions could also provide communications to the poor in those regions. Computer, software, and peripheral equipment makers should provide their products on the same basis, as should service organizations with training and other services.

To help motivate all commercial providers, the governments of industrial nations should change tax laws to provide exemptions, credits, and other incentives to companies that take such actions. Donating companies would get good publicity in both the industrial and the developing world and they would help create new markets. These companies might also consider obtaining "promissory notes" from the beneficiary countries, in the form of information work to be delivered in the future; Sri Lankan engineers could wind up working for Intel from their home base, or in person at the company's plants.

3. Foundations and wealthy individuals: help the poor access the Information Marketplace.

Local and national benefactors, be they individuals or foundations, should establish programs that would help poor people join the Information Marketplace in a productive way. Teaching people how to leverage the Information Marketplace is best done through local projects, and such projects are most effectively run by specific foundations. Already, for example, the Soros Foundation, established by financier George Soros, has made available funds through its Internet and other programs to bring former Eastern Bloc citizens into the new world of information. The Electronic Frontier Foundation, established by Lotus founder Mitch Kapor, strives to ensure accessibility of information and information services to all U.S. citizens on an equitable and affordable basis, while focusing on protecting the principles embodied in the U.S. Constitution. There is room for many more such actions to address the needs of people in developing nations and the urban and rural poor in the industrialized world. Such contributions can make a big difference in closing the rich-poor gap!

4. The poor: seek to benefit from the New World.

To a person or nation that lacks food and fundamental infrastructure like roads and municipal water systems, the prospects of improvement via the Information Marketplace may sound naive, unreachable, even silly. While there is some justification for these perceptions, the Information Marketplace is still important as an economic lever for the poor, because they can use it to learn new skills and to sell their work. As the more entrepreneurial of these people acquire a new skill or identify a useful service they can offer, their example will be emulated by their more reticent peers, sweeping an increasingly larger fraction of their population up the economic ladder.

Such a rise, of course, will not be possible if poor people decide that the Information Marketplace is not for them, or that it is an instrument by which the rich can exploit the poor. Wealthy nations should vigilantly look for and undermine schemes that exploit the poor, such as the charging of excessively high fees by information work brokers. They should also try to promote the Information Marketplace's value to the poor, and entice the poor to embrace it, by focusing on truly useful projects that improve local agriculture, health care, and education.

Meanwhile, poor people should actively prepare to join the Information Marketplace by enabling and participating in projects to that end, by helping their peers toward the same goal, by actively voicing their wishes and concerns, and by being willing to become involved. With help from the rich and some good luck, the world may be able to stop the rich-poor gap from growing, and may perhaps even shrink it.

Business Agenda

Businesses—and all organizations—stand to benefit the most, and soonest, from the Information Marketplace. Smart use of the strategies below will enable companies to increase efficiency, improve customer service, open new markets, and grab a big piece of the $3 trillion in trade that will take place over the Information Marketplace early in the twenty-first century.

1. Plunge in and experiment, but don't overmanage or monopolize.

The world is full of organizations that sit on the sidelines, talking about business on the Internet and the Web while waiting for somebody else to "pull the snake out of the hole," as we say in my native Greece. I ran into a major U.S. toy manufacturer doing just that. The company decided to stay out of the Internet because it concluded that children will not be interested in information about toys but rather in the toys themselves. The assertion about children's tastes is accurate enough, but the conclusion is dead wrong, because the Information Marketplace can deliver fantastic toys—not only information about toys. Just look at the game Horse Race in chapter 2 and all the many games we discuss in chapter 6.

In many meetings with business executives I hear the recurring mantra: "You don't understand. We must show profit during this quarter. We can't afford to go into this without a visible ROI (return on investment)." To which my knee-jerk answer is "If a ROI is too visible, it may be too late." This is a time to explore new territories. You must find them, ahead of others, if you want to win big. As my colleague economist Lester Thurow asks, "What was the ROI for Columbus's exploration?"

Focus on establishing a process that reveals any potential ROI . . . ahead of your competitors. You don't have to go into this in a big and costly way. It is easy and inexpensive to experiment. Let some of your younger employees wander and play around on the Web, trying out their ideas and observing other people's ideas. In doing so, these young explorers will be mixing their knowledge of your company with the new world of information, while staying vigilant of any new opportunities that may jump out of nowhere in front of their eyes. As you do so, remember not to overmanage these experiments. Granted it is difficult to manage purposefully in the presence of so many unknowns. So don't! Instead, let a small part of your organization experiment with a big part of your imagination.

If your organization is already on the Net, don't think for a moment that you can own or control a large part of this new world, even if your company is the largest in the world. If you doubt this advice, take a look at the war of the spiders and the battle of the pipes in chapter 2. Businesses trying to monopolize even a piece of the Information Marketplace can't succeed, any more than an organization striving to own all the factories, shops, and service establishments of the world. Besides, there is plenty of money to be made for everyone. Instead, you should support the Information Marketplace, working with your government and your competitors, to ensure that an effective information infrastructure is built that you (and your competitors) can use to improve your own operations and reach a far bigger market than you do today.

2. Use knowledge of your business, common sense, and avoid hype.

Hold off on appointing someone as your organization's chief knowledge officer. Fire any consultant who advises you on the bandwidth of your network instead of on what it might do for your company. Don't become impressed with slogans like "the instant organization" or "the knowledge-empowered worker." The blinding flash created by these seductive terms hides what, if anything, they might do to improve your company. Use your detailed knowledge of your business and focus on the specific steps that will help your organization attain concrete goals. Before you leave the experimental stage and commit to a new initiative in the Information Marketplace, ask the obvious and tough questions: How will your employees use it? How will clients use it" What tangible benefits will each experience? How much will it cost? Strip away the obfuscating techno-babble, the irrational fears, and all the rotten layers of the onion until you find the core. If what is left makes sense, try it. You probably have a winner.

As a business leader or entrepreneur, search for and create the new applications of the Information Marketplace that will serve human needs, starting with the many kinds of goods and services described in part II of this book. Most important, ask whether what you do today, or what you plan to do in the future, could be done better, faster, more accurately, or more cheaply by using the new forces and capabilities of the Information Age, as we detail in items 3, 4, 5, and 6, below. And carry out these steps across the full life cycle of your product or service—concept generation, research, design, development, production, marketing, advertising, sales, maintenance, service, and retirement.

3. Step 1: exploit electronic bulldozers.

You can improve almost any aspect of your operations by asking a simple question: "What information work being carried out by our people today can be offloaded onto machines?" Look at routine people-to-people and people-to-machine transactions and probe whether they can be converted in part or whole to machine-machine transactions. Remember the e-form of chapter 4, filled by speaking to your machine "Computer take us to Athens this weekend." It took three seconds to say, and it caused your machine and the airline to work for ten minutes, resulting in a 20,000 percent efficiency gain. Ask, too, where you can use "data sockets" (chapter 5)—e-forms posted on a designated computer site and updated at pre-agreed intervals—to document, for example, the sales orders received by a company's subsidiaries on a daily, weekly, or monthly basis. Employees can then count on this up-to-date information, "plug into it" with their own computers, and use it for any purpose they see fit.

Devising many of these work-saving electronic bulldozers does not require a knowledge of high technology, but rather human agreements on what information they should contain and where they should be applied to do the most good. E-forms and data sockets are most useful when developed by the employees who will use them. Let employees apply their creativity and knowledge of their part of the organization to devise the procedures. The results will inevitably be more effective than anything invented by a central executive authority.

Finally, seek agreements with other organizations in your industry on common e-forms and information standards that will improve everyone's operations. By extending this mindset to your common-interest group and by joining your competitors in devising the right e-forms and data sockets, you will make your, and their, businesses more convenient, more accessible, and less costly to clients and suppliers.

4. Step 2: exploit electronic proximity.

Ask yourself what fruits you could reap if physically distant parts of your organization could be brought closer together. The sites may be thousands of miles apart—or down the corridor. Electronic proximity through better e-mail, telework, and groupwork (chapter 4) can make a difference. The key question to ask is whether, compared to its physical counterpart, an effective working relationship can be maintained over the Information Marketplace. Where it cannot, benefits may still be had by breaking down the work into parts that can be handled effectively via the Information Marketplace and those that require physical proximity. Even through partial measures, you may be able to accelerate cooperative work processes and decrease the need for physical travel. As the expert teams at British Petroleum have discovered (chapter 9), the payoff can be substantial. To that end, consider building work centers for your organization's employees in low-cost, high-quality-of-life parts of the world. Or, if you are an entrepreneur contemplating new businesses, consider building such work centers for use by anyone working remotely for any organization. As you move forward, though, remember the forces of the cave. Employees who work together via the Net need to meet face-to-face in order to build the kind of trust essential to human teamwork.

5. **Step 3: explore conversational speech systems.**

If your organization communicates with clients or suppliers in a narrow domain, such as catalog orders or queries about hotel rates, store hours, or driving directions, you might be one of the early users of speech-understanding systems that can improve service while dramatically reducing your costs. You know you have a narrow domain if the kinds of questions being asked are in the tens or few hundreds, even if they can be asked in thousands of different ways. Like the airline reservation system discussed in chapter 3, when the conversation is tightly focused these systems can exhibit 90 percent or better comprehension. Since they can be made to paraphrase and repeat what they have heard, they can be corrected on the first try, raising accuracy even further. For customers, these systems are much more human, and effective, than today's automated phone operators that torture us with their endless litanies of "if you want x then press y." Look for a wide proliferation of these natural interfaces and then look again before you decide that they may not work for your purposes.

6. **Step 4: integrate human and machine processes.**

Besides trying to exploit the two major forces of electronic bulldozers and electronic proximity and the spoken dialog interfaces, you should review whether you can profitably use other tools and capabilities of the Information Age. Look at mass individualization, discussed in chapter 5, through which products and services can be tailored automatically to different individuals' needs. Visit reverse advertising, where clients specify their wishes and you (and your competitors) respond if you can meet them. And examine the advanced e-mail tools, security systems, pipe managers, and hyperorganizers discussed in chapter 4 to see if they make sense for your business.

Once you have identified the electronic bulldozers, electronic proximity techniques, spoken language interfaces, customization, and other information tools that might help your company, you should integrate them within your organizational processes and dovetail them with the employees who use them. This is the most important step, because organizations will increasingly be only as good as their combined human and machine processes. And the companies that manage to create such fine-tuned processes will be the ones that excel.

7. Don't cut spending at the expense of your customers.

I wish this obvious admonition were unnecessary. But as I have repeatedly remarked, organizations keep installing automated systems that save them money at the expense of the rest of us. The automated telephone operators that force us to press "1" for this and "2" for that may cost a company only a few thousand dollars and save it a few thousand more a year in reduced personnel expenditures. But the thousands of people a year who call this auto-moron waste millions of dollars of their time, get poorer service—and grow disgusted with your organization. If you engage in such practices, stop. Your clients may endure these assaults for a while, but sooner or later they will leave you, as they should, and go to your competitors who provide good customer service rather than torture in the form of long delays while pretending to care with the cheap and dishonest repetition of "Please hold on. Your call is important to us."

8. Look for new opportunities.

In 1996, perhaps $200 million changed hands over the Internet. In 1997, the figure was over $1 billion. By the turn of the century, current forecasts place it at around $10 billion. Some twenty to thirty years into the twenty-first century, the figure may reach $2 to $3 trillion. Looking for new opportunities seems worthwhile. I have been overwhelmed by people asking me, "What companies should we invest in?" or "What kind of company should I start to take advantage of the Internet?" There is no magician who can provide the reply. Indeed, the challenge before the corporate world and aspiring entrepreneurs is to answer these questions.

Nonetheless, there is some early evidence about the types of businesses that seem to "take" on the Internet. In the late 1990s, Dell Computer was selling $1 million a day of personal computer systems consisting of components selected by clients over a friendly Internet site. Amazon.com was selling about $1 million a week of books from a list of 2.5 million titles; its site offered reviews, bestseller lists, and even the comments of readers about certain books. Similar businesses were selling hi-fi equipment, CDs, and many more products, while others were offering services, such as online stock trading, and selling new cars directly from manufacturers at the lowest possible price.

A large number of new businesses were on the drawing board as this book went to press. Many were in the category of intermediaries, like real estate, insurance, and employment brokers. One organization was giving away its employment matching service to qualified job seekers who filled out the company's résumé forms, but charged dearly the potential employers who searched through the million or so accumulated résumés for potential employees. Even familiar activities like buying groceries or finding a good used car were vigorously pursued online. As in any new business endeavor, the possible payoffs are offset by equally high risks. For example, if you wanted to go into retail food distribution based on orders placed online, you might provide a delivery service that shops from well-known and well-stocked grocery stores specified by your client, as is done today by several companies. Or, you might opt for a new grocery store that will cater exclusively to online orders and therefore can be housed in inferior but much cheaper real estate that no one will ever visit. This option requires a high investment and carries a high risk if it fails. On the other hand, it may become a huge success because of the convenience and low costs it would offer consumers.

Stretch your brain for new opportunities and page through part II once again. The many commercial services imagined there may soon be reality, and your company can be among the first to provide them.

Techie Agenda

The computer has already secured its place in history as one of the most powerful tools ever created. Yet, PCs and software systems are confusing and barely usable. Unless we, the techies of the world, seriously tackle this problem, the machines may become more effective as doorstops and the entire Information Revolution may be remembered as a fanciful fad pursued by people who enjoy sparring with complex machines! It's high time to make computers do what we want done. Fortune and fame await the hardware architects, software designers, and communications engineers who can meet any of the challenges below.

1. Make machines and programs much easier to use.

Remember that all the tools and systems you invent will be used by people, so bring these artifacts into our lives, not the other way around. In ergonomics, designers strive to match the physical aspects of machines to the human form. System designers of the Information Marketplace must adopt a similar mindset and discipline, perhaps called "humanomics," in which they match tomorrow's information systems to natural human capabilities, habits, and needs—both physical and mental. Here are some places to start:

1. *Reverse a system-design point of view,* rooted in decades-old habits, that throws control of subsystems like networks, modems, and computers on the user—the complexity fault of chapter 12. Don't make the nurse have to program the communications setting for her modem. Design systems that do this for her when she asks for the patient's records from a remote hospital, regardless of network and software calamities encountered in the process. In other words, give the user a gas pedal to control the car—not just ignition and fuel-air mixture controls that must be expertly juggled to control speed.

2. *Tailor systems to human specialties.* It's time to move away from developing general-purpose systems like word processors, spreadsheets, databases, compilers, and the like that are as useful to accountants as they are to musicians. We've milked generality enough, and we'll no doubt continue to use it across specialties, but now we must focus on developing systems that cater to the specific needs of accountants, musicians, doctors, and other professionals. As we discuss in chapter 12, this will allow users to delve into familiar concepts and approaches of their speciality and therefore to be more comfortable and more productive.

3. *Pursue customizable software and software that can grow.* In the spirit of gentle slope systems (chapter 12), develop powerful modules and composition schemes that can be used with little effort by normal people, like putting together a model train, to render increasingly more useful function for small increments of added work. And develop individually tailored software systems that can grow with time in an orderly way, without losing their nimbleness, so as to avoid the ratchet fault discussed in chapter 12.

4. *Say good-bye to structure and hello to meaning.* Quit working so hard on the structure of information and develop systems centered on its meaning. This very tall order calls for changing individual computer systems and, as we discuss below, information infrastructures so the meaning of information is

represented and handled with as much ease as its structure. For example, today's e-mail systems can't "understand" the information contained in subject headers. That task falls to the human, who wastes time reading e-mail she's not interested in or doesn't have to act on right away. In a meaning-oriented e-mail system, the headers would be understood, by agreement of the participants, to fall into preestablished categories, a sort of Dewey decimal system for the actively discussed business on hand, with the requisite escape for free text headers in the absence of a suitable category. Knowing something about the meaning of messages would enable automatic filters to route them and file them, thereby saving much human time. Techies averse to centralized taxonomies of meaning need not despair, as we discuss next.

2. Advance the Internet and Web into a full-fledged information infrastructure.

The information infrastructure provided by today's Internet and Web is distributed, data oriented, heavy in its demands for human attention, and oblivious to the meaning of the information flowing over it. Some obvious improvements, in the area of security and payments, for example, are already under way. That's not enough. If it is to evolve into the robust Information Marketplace described in this book, it needs additional capabilities, beginning with the following:

1. *Better performance.* The number of users and network nodes continues to grow dramatically. Today's Internet/Web accommodates only some 40 million users, who often experience a variety of problems in locating servers and successfully sending and receiving information. Speeds are slow and delays long. Work simultaneously on adjusting the Internet, the Web, and individual computer software to overcome these problems.

2. *Function and tool orientation.* When we click on a distant site, we should be able to access not just pictures, sounds, text, and video but programs that carry out useful actions, like a new text editor or a fancy game. We should also be able to combine such programs as if they were pieces of text being cut and pasted on one machine. Multiplatform languages like Java are a step in this direction, but much more must be done to make the

Internet/Web as centered on actions as it is on viewing passive data. Develop approaches to that end.

3. *Automatization.* Develop e-forms, data sockets, and other electronic bulldozer innovations so that users don't tire their brains and eyeballs poring over scores of Web pages and e-mail messages. Create automatization routines that offload low-level human brainwork onto machines.

4. *Shared representations of meaning.* It's unlikely that anyone has or should develop a workable "dictionary" that assigns an absolute and unique meaning to the various words we expect computers to understand. Instead, give specialty users such as radiologists the ability to attach meaning to terms and symbols shared only by them and their computers. Eventually, common terminology—and the ability to recognize it—will evolve across disciplines, making it possible to grow the information infrastructure toward broader and more universally shared concepts. Stay focused on this goal of establishing shared concepts among machines, which is for the Information Marketplace as important as the acquisition of language among humans. Meeting this goal is the secret to achieving the automatization, greater intelligence, and added utility that will make the information infrastructure truly useful.

3. Develop software for the Information Age.

In chapter 4 we discussed some new kinds of shared software—for finding and organizing information, transporting information more effectively, automating machine-to-machine transactions that offload human brainwork, making possible groupwork and telework, and so on. Groupwork alone will require thousands of different software packages to handle various kinds of meetings—one-on-one, small group, workshop, and theater—for different professionals, from ophthalmologists to steel makers. Many more thousands of application software packages, both commercial and internal-to-an-organization, will be needed. Go after these new software packages. Work with the entrepreneurs and business executives who are trying to discover what they should be.

Make sure the new tools you develop do not promote the misuses explained in chapter 12. Also, keep your eye on the compass; your tools should strive to increase human productivity as dramatically as industrialization did. This software will constitute the primary way people experience and use the Information Marketplace and it will take decades to develop. You are in the enviable and privileged position of providing, through thousands of new and useful programs, the true utility of the Information Marketplace to your fellow human beings and their descendants.

4. Invent system facades that surpass operating systems and browsers.

By a "facade" I mean the totality of interfaces people encounter when using a computer system. We use operating system facades like Windows to control the information in our computer, and browsers like Netscape to find and manipulate information on the Internet/Web. New facades will be needed to help us work in a uniform and natural way with our own information and with the information that belongs to others throughout the Information Marketplace. A naive blending of the two old forms—operating systems and browsers—will not suffice, since they are sufficiently different to prevent uniformity and since the new facades will have to deal with the many new capabilities that stand between today's Internet/Web and tomorrow's Information Marketplace. To grab this brass ring, you should invent a metaphor as simple, unique, and fresh as the WIMP (Windows, Icons, Menus, Pointing) interface was when it was first introduced, but which is suitable for the new world of information. It should not favor either operating systems or browser functions, and it should be able to deal naturally, simply, and efficiently with the desirable new capabilities of the Web—a function-and-tool orientation, automatization capabilities, and dealing with the meaning of information.

5. Improve human-machine interfaces.

Better interfaces are needed, too. The ideal human-machine interfaces convey concepts between people and machine in the most natural way. Maybe that's via a keyboard or mouse, but maybe that's a speech system, a haptic glove, or an eyeball tracker. The interfaces need not be the same for input and output—speech may be better suited for input and vision for output. Nor do they need to be the same across different uses. You have a rich world to explore and many interfaces to invent.

Explore and invent the interfaces that best convey different classes of concepts by studying what people do and how they do it. Also explore the possibility of moving these interfaces to a higher level, closer to what their users want to do. Do not isolate, for example, the speech-understanding interface from the airline reservation system program, but combine the two in one interface program that better serves users' needs.

Finally, consider that all the different interfaces developed so far, regardless of whether they use vision, speech, gestures, or touch, seem to be surface variations of something deeper and more constant—perhaps an inherent linguistic or thought process we have used for centuries to convey concepts among ourselves. You should probe for such a powerful new organizing principle that may unify human-machine interfaces and help convey concepts more effectively in both directions.

6. Pursue high-risk, high-payoff research.

The practical development that thousands of techies undertake should be counterbalanced with some daring experimentation. Today's crazy idea could well be tomorrow's routine. If you are a researcher so inclined, this is advice for you.

One of the most exciting prospects for future computer systems is their potential ability to mimic human intelligence and therefore serve us like intelligent assistants. This is not possible today, except in highly specialized situations (narrow domains) such as booking airline reservations or asking about the weather. Of the various options to make machines intelligent, machine learning is probably the most likely to create a new revolution. If machines could learn from experience rather than by being explicitly programmed, they would be able to grow in knowledge and usefulness with time.

The problem is extremely difficult, and has not been cracked despite massive efforts over four decades. That doesn't make it unsolvable—just tough. The machine-learning problem, and the broader artificial intelligence (AI) problem of understanding human intelligence and making machines behave intelligently, should be vigorously pursued in the laboratory. Today, many AI researchers are timid about doing so and couch all kinds of other problems under the AI rubric because underperformance and overpromises have left the world jaded and the research funders nervous. Never mind. Boldly go after the big AI dreams. The potential payoffs justify the risks.

Major scientific breakthroughs are needed in other vital and challenging areas, such as making computers with biological materials and discovering the computer science that seems to underlie biological processes, developing effective machine vision, inventing a new and useful theory of information beyond Claude Shannon's classic, understanding the economics of information, and, for the very daring, searching for a unified theory, based on information, that underlies the natural sciences. These and other ambitious goals should all be vigorously pursued. With a growing panoply of lab results and a better understanding of computer science and technology, the Information Age will be on solid footing for continued progress throughout the entire twenty-first century.

7. Accept and integrate the humie within you.

Remember that the techie and humie are artificial distinctions created by the Enlightenment's split of faith from reason three hundred years ago. Don't put up with this narrow view, for you will be unable to practically cope with an increasingly complex world that does not neatly divide into these two bins. Awaken and embrace the humie within you. Your technical directions will then be better guided by human need. You may even become a better leader—it's about time techies graduated from being in the "servile arts" to people who can combine lofty humanistic ideas and challenges with a deep technological understanding of the world—whether you become in charge of a project, an organization, or a nation.

Humie Agenda

If you are convinced that everything humanistic stands above all that is technological, you are laboring under a profound misunderstanding, for technology is a child of humanity and an integral part of it. Rather than fight or condemn what will be, consider leading the world into its next age by applying your deep understanding of human nature. You are in a unique position to shape the future. Consider beginning with the steps discussed here.

1. Conceive a human ideology for the Information Age.

Frame a manifesto that will help society understand how to behave in the new world of information.

Take, for example, the issue of intrusion: Does electronic proximity give people the right to penetrate your world with their e-mail? Are you obliged to read every message you get, or can you ignore whatever messages you want—without guilt? I would opt for the latter. Should telemarketers identify themselves explicitly, or is it fair game for them to try and parade as acquaintances as they often do today? Remember, you may get 1,200 telemarketing e-mail messages a day. I would want them to clearly identify themselves when they knock at my electronic door.

Or take privacy: Should people have an obligation to check with you before using information concerning you, or should information be regarded as a "free" good if our society is to be free? What of anonymity? Do people have a right to it? I don't think so. I would require people to digitally sign anything they want me to access. This would not prevent them from sending anonymous messages to those who are willing to accept them. But a commonly accepted identification and signing scheme should exist in the Information Marketplace for those who want to use it. And people should be able to keep their doors closed to anything and anyone that is not identified.

What about intellectual property rights? Should people expect some protection for the information they create, or once it is on the Net should it be available to all, again in the interest of "freedom"? Some actions are already taking place to enable copyrighting of information on the Information Marketplace. Unfortunately, this is where most people stop debating the issue of intellectual property rights, because the only "content" they can imagine is copyrightable material like text, pictures, and video—the stuff of multimedia. They ignore something ten times bigger—the buying and selling of information work (chapter 11). If, for example, a skilled doctor diagnoses people over the network, should some enterprising person be allowed to collect her diagnostic procedures and those of other doctors and use them subsequently as a basis for an automated, case-based "doctor" database, without the practitioners' permission? Certainly not in my manifesto! Humans have always had the right to their own work. We should not allow freeloading cutely disguised as the "freedom" to use information without restraints. People should not be able to capitalize on others' work, without their explicit permission.

The manifesto you develop should help us to understand many more issues about how we wish to handle information work in our society—exploitation of human telework, liability for work performed by an anonymous computer program, trade in information labor—both nationally and internationally—with the attendant need or not for taxes and tariffs. We need insight on the balance of citizen and government privacy rights (chapter 10), the tradeoff between tribal and diverse cultural imperatives (chapter 13), and more. The world needs your experience in framing and answering the complex psychological, aesthetic, philosophical, and ethical questions new technology raises.

2. Identify and pursue worthy projects with human goals.

Instead of insisting that technology dehumanizes the world, try to dream up new projects that use technology to benefit humanity, like the Virtual Compassion Corps (chapter 5), which matches the providers of human help with those who could use it. Or the histori-copter of chapter 6 and the World Heritage Library of chapter 8, through which anyone can travel through space and time to explore the artifacts and principal events of our history.

Then do your share to help turn these ideas into reality. Don't shy away just because you consider yourself an "idea" person rather than a techie. In the Information Marketplace, ideas, especially good ones, can be easily converted into major actions. Just look at the Web itself: Tim Berners-Lee started it, essentially single-handedly, as a way for scientists to share documents, and it has mushroomed into an unstoppable global force.

3. Warn the world of emerging problems.

Watch for ways in which the new information technologies and ancient human nature may clash. When you detect the onset of trouble, speak up and let the technologists know, so they can take action. Your knowledge of history should be very helpful in this re-gard, since no doubt you will have seen some of the same problems develop in the past under a different guise. Your vigilance is crucial, because the Information Marketplace is expected to continue to evolve for most of the twenty-first century and in doing so will give rise to lots of problems and difficulties that may go otherwise un-detected until it's too late.

4. Debate the humanistic aspects of the Information Marketplace.

In chapter 14 I declare that the Information Marketplace will ex-acerbate the unnatural split between the techie and humie within us, and will lead us to unify the two halves in order to cope with an increasingly complex world where techie and humie issues will be even more intertwined. Throughout the book I have made several counterintuitive forecasts:

- The Information Marketplace will simultaneously help diversity and tribalism around the world.

- It will reunify divided peoples and may even lead to a redefinition of nations from geographic entities to electronic networks that connect indigenous people wherever they may be.

- It will not lead to a single global culture but rather to a thin veneer of shared norms.

- The balance between good and evil will not change and no major new laws will be needed, since human nature will remain unchanged.

- The economic value of information work is indistinguishable from the economic value of physical work; and

- the value of most information is determined by the value of the human desires it helps to satisfy.

- Left to its own devices, the Information Marketplace will increase the gap between the rich and poor nations and rich and poor people.

- We'll face the psychological tensions of the urban villager—half urban sophisticate roaming the world's networks for professional purposes and half villager meeting the same people every day while working at home and catering to family and friends.

- Human emotions will pass only partially through the new medium; and

- the primal human emotions—the forces of the cave—so crucial to our everyday lives will not pass at all.

Are these ideas bunk? I encourage you to debate these points and many others that I did not have the foresight to raise. By discussing them you will help bring attention and clarity to them and you will help prepare all of us to enter the new world of information.

5. Accept and integrate the techie within you.

As I suggested above to the techies of the world, don't put up with the artificial split between techies and humies that has prevailed since the Enlightenment. If you don't abandon this old prejudice, you'll fail to understand an increasingly complex world whose problems and opportunities combine the techie and the humie. Try to wake up and understand the techie within you so that it may link arms with your humie side and make you whole. And if you have them, please abandon your stubborn beliefs that technology can be purchased or fashioned at the service of an independent and far loftier human purpose. Try to see technology for what it is—a child of humanity, an integrated part of us all.

Humies (and techies) are not the only ones to help bridge the humie-techie split. Parents: ensure that your children grow up without this humie-techie hang-up. Set an example, and encourage them to explore and integrate the techie and humie within them. Have them study science and art, play soccer and chess, talk about engines and emotions. Educators: revise your grade school, high school, and university curricula to the same end. You need not give everybody an equal dose of techie and humie skills, but you do need to ensure that everyone is given examples of both, a chance to cultivate the techie and humie within them and, more important, the ability to integrate both in their own unique mixture.

Government Agenda

As a citizen who will be greatly affected by the new world of information, you want to know what your government should do with the Information Marketplace and in what directions you might try to steer its actions. As a government leader or politician, you no doubt perceive the political and economic significance of the Information Marketplace. That's clear enough from the wealth of ministerial pronouncements and governmental declarations about information technology, worldwide. What's less clear to the governing and governed alike is the role government should play and the goals it should set for deployment and use of this new medium in your nation, especially in light of its complicated nature—part national infrastructure and part international resource; part commercially oriented and part a free common good; part controllable by you and part dependent on others who are many physical and cultural miles away. You correctly sense that if government can somehow help the Information Marketplace grow, your nation will be all the stronger. But how? You may begin with the following:

    1. **Don't under-regulate or over-regulate the Information Marketplace.**

That's more easily said than done. In the United States, which has one of the world's least intrusive governments, there is a perpetual fear of government involvement in almost any activity, lest it lead to more taxes and greater inefficiencies. Ironically, the anti–government-involvement mantra is mouthed most loudly by politicians. That makes for good press, but it is contrary to government's obligation and sole reason for existence: to anticipate, plan, and carry out policies and actions for the common good. Greater thought must be given to increased governmental involvement at the infrastructure level, despite the knee-jerk tendency in the United States and most of the West to under-regulate. The medium is too richly linked to the public interest to be left entirely to market and technological forces.

Other governments are all too eager to over-regulate. First they strive to keep all undesirable foreign elements from crossing their borders through the Internet. That is exceedingly difficult. Then they assure their citizens that they can provide most of the necessary online services by regulating the overall network and controlling who and what goes on it. That is impossible. As this book goes to press, a few governments in Europe have gone through the provide-all-services charade, and some in Asia are going through the keep-people-out stage. If you are a leader of such a nation, give up on the notion that your government can own and control the Information Marketplace. As I stated in the Business Agenda, it can't any more than it can "own" control of your nation's shops, manufacturing, and service establishments.

The Information Marketplace has many characteristics that concern the common welfare of a nation's people, like the privacy of information, preservation of culture, handling of tribal clashes within and across national borders, rights to information and information work, taxes and tariffs, and more. With its legislative powers, your government may want to address many of these critical issues, and anticipate or correct mishaps that may develop, especially in the privacy, authentication, and payment areas. You should also develop policies that help close the rich-poor gap within your nation, such as the tax-credits for industry that we discussed earlier.

2. Help the information infrastructure grow, and help people use it.

Without a solid information infrastructure there can be no Information Marketplace in your nation. If you want your people to partake of its benefits, you must ensure that your infrastructure can grow and be connected to the global infrastructure. Regardless of whether your telecommunications provider is a government monopoly or an assembly of deregulated competitors, you should ensure that most homes and businesses in your nation will become interconnected soon, and with a decent quality of service.

U.S. leaders: keep in mind that since you are opposed to pursuing such tasks centrally, your nation is at a disadvantage relative to others like Japan, which can decide, as it has, to lop off $300 billion from its tax base to provide glass fiber to every home and office. Look carefully at the major information carriers. Most talk expansively about the future but are reluctant to make big infrastructure investments because they do not see a sufficient payback. Historically, the U.S. government has always helped with the formation of major infrastructures like highways and telephone systems. You should rethink your hands-off policy, however fashionable, and courageously explore whether it would be in the common interest of the American people to do for the information infrastructure something similar to the Highway Act. I think it would, for example, in providing the necessary capital to ensure high-speed connectivity among all U.S. homes and offices within a decade.

Regardless of whether or not your nation is like the United States, your government should ensure that commerce, learning, health, entertainment, and other goals can be pursued over the Information Marketplace in alignment with your nation's cultural norms. When feasible, your government should also establish policies for privacy, and for avoiding info-junk and the many other dangers discussed throughout this book.

As a promoter of progress, your government should support and experiment with new approaches to learning. But as a protector of people, it should guard against theatrically appealing but unproven massive deployments of technology toward education (chapter 8). The same applies to health care. Ensure that hospitals and government policies allow information to be used effectively but properly in this important arena. Meanwhile, pursue new initiatives that could make a big difference to the less fortunate in your society, like tax incentives for organizations that donate goods and services to the poor.

3. Preserve local culture and strike international agreements.

National leaders: your people and their ancestors have worked hard, for hundreds if not thousands of years, to establish your nation's indigenous culture or cultures. You are proud of this and so are they. So naturally, when you hear people lecture you about how you should drop all these old ways and adopt a universal new culture (read a U.S. culture) since you are now interconnected with the rest of the world, you cringe. Your reaction is justified: don't give in to such suggestions. As we discussed in chapter 13, use every political, technical, and other resource in your nation to ensure that the culture your ancestors built is preserved.

Meanwhile, begin exploiting electronic proximity to extend your nation beyond its landmass boundaries to an ethnic network that includes your people abroad. Then work on extending the definition of your "nation" from a geographic landmass to mean this broader network. Like a school's alumni, all those people have a vested interest in your nation, and have a great deal to contribute to it, too.

At this point, politicians often say to me, "Maybe we can manage matters within our borders, but how can we handle cross-border information clashes over which we have no jurisdiction?" Fortunately, you don't have to reinvent the wheel. Procedures are already established to control cross-border flows of physical pornographic materials and drugs, international crime violations, international trade problems, and misdeeds in ambiguous jurisdictions such as the high seas. With these precedents, sit down with your counterparts in other nations and strike new international agreements that determine how you will handle information transgressions that cross your respective borders.

Naturally, a group of nations, perhaps trading partners, may see such issues similarly, and adopt shared policies. Other nations will not see eye-to-eye. Either way, such international agreements will be difficult and time consuming. But they must be carried out, because international crimes on the Information Marketplace will outnumber current cross-border crimes as a consequence of electronic proximity. This means, too, that nations will have to take far more seriously international approaches for resolving conflicts than they have to date.

4. Streamline your government's handling of information.

Government is a huge organization. Officials should apply much of what is presented in the Business Agenda to processes within government. Consider how your government might automate its transactions to offload low-level brainwork, how it might exploit electronic proximity to link key parts of your government, and how it might make the work of government employees faster, simpler, more accurate, and less expensive. Your military and intelligence establishments can also use the Information Marketplace to reduce expenses and the need for physical weapons through more extensive use of simulation. And don't forget your government's massive purchases, which can also be dramatically improved using e-forms and data sockets. Since a lot of what you will be ultimately able to do is not yet visible, consider establishing your national governmental intranet—a gigantic information infrastructure linking all of your nation's legislative, judicial, and executive agencies. This governmental Information Marketplace should make possible any future intergovernmental transactions as the need for them arises. Consider, too, extending the information reach of your government internationally. An international governmental intranet will evolve naturally if not speedily and will supplement and support the established diplomatic, trade, business, and law-enforcement channels for intergovernmental discussions and transactions.

Equally exciting is what your government can achieve in its interactions with your citizenry. Polls, voting sessions, and comment solicitations from the people are the obvious first steps, as are broadcasts from the government to the people. However, you can go further. Your government should consider automating the filing of tax returns, and many other processes required of the citizenry, such as obtaining licenses, affidavits, duplicates, and other documents for which they must now stand in line. It will save them, and you, time and money.

Your government's "customer service" can be improved by offering vital information to people interested in laws, taxation, regulations, requests for proposals, and other data about their government that affects them directly. Make it easy for them to find out what they want. As security systems improve, the scope of such inquiries will extend to personal information, and will permit your government to provide a wider range of customized responses that replace impersonal and ineffective boilerplate replies.

There will also be many things that cannot be done with the new information technologies. I have discussed in chapter 10 why electronic town halls among many people are impossible, and why polls cannot be used excessively, lest they replace government with mob rule. Whether as a citizen or politician, educate yourself on which procedures work and why.

5. **Non-democratic governments: get early credit toward democratization.**

If you are a leader of a non-democratic country, realize that if you want to participate in the global economic arena, you and your people will be subjected to new democratization forces. I am not focusing here on the often touted but questionable premise that the Internet will cross your national borders regardless of what you do. As we discussed in the Big Brother section of chapter 13, a dictator could keep his country isolated from the global Information Marketplace and even use the new information technologies to strengthen his totalitarian grip. But if you want your people and industries to effectively sell their labor or their wares, or to disseminate and collect information, or to promote international learning, or pursue any of the hundreds of other uses of the global Information Marketplace, you will necessarily have to play by that medium's rules of engagement, . . . which are made by the predominantly democratic countries that established it.

If you choose to join the global Information Marketplace, you will open a huge door through which many of your people can step. It will be increasingly difficult for your government to monitor all that they are doing. Their casual exchanges on the Information Marketplace alone will expose them to democratization forces. As recent history suggests, wanting to live like your richer, more democratic neighbors is a force big enough to topple protective walls and superpowers.

Exposure to new technology does not mean democracy is inevitable in your nation. That is a naïve assertion. But it will stare you in the face if you choose to play in the world economy via the global Information Marketplace. Since I have yet to meet the national leader who does not want this economic benefit for his people, you might as well accelerate the democratization of your nation, and get early credit from your people and from the world for doing so voluntarily, independently, and with the foresight of a true leader.

Appendix
The Five Pillars of the Information Age

At their simplest level, all the fascinating activities that will take place in the Information Marketplace are made possible by the generating, processing, transmitting, and receiving of information. As we stated in chapter 2, this all rests on five essential pillars:

1. Numbers are used to represent all information.
2. These numbers are expressed with 1s and 0s.
3. Computers transform information by doing arithmetic on these numbers.
4. Communications systems move information around by moving these numbers.
5. Computers and communications systems combine to form computer networks—the basis of tomorrow's information infrastructures—which in turn are the basis of the Information Marketplace.

Understanding these pillars is equivalent to knowing that an engine, tires, fuel, and a chassis are needed to make a car. But to really appreciate how a car works—and how you can leverage it—you'd like to know how the engine converts fuel into power that turns tires that make the chassis move and take you where you want to go. This appendix provides a straightforward explanation of the five pillars along with some of the technical terms and concepts that we increasingly bump into every day, such as *pixel, sensor, firewalls, chip, bandwidth, compression, Internet address, LAN* and *WAN, MIPS* and *MHz, digital* and *analog,* and more.

Let's consider what actually happens when you, happily planning your next trip to Asia, turn to your computer and ask it to get the weather forecast for Hong Kong.

1. Numbers, Numbers Everywhere

As you address your computer, a microphone picks up the sound impulses of your voice, "What's the weather like in Hong Kong?" and your computer converts them to a list of numbers. Recognizing the pattern of these numbers as a question about the weather, your computer then issues a "get Hong Kong weather" command—a different list of numbers that it has been told is understood by a certain machine in Hong Kong. Your computer sends the string of numbers to the Hong Kong machine via telephone lines. The Hong Kong machine receives the numbers, deciphers the question, and ships the forecast back to your computer as yet another string of numbers. Your computer converts these numbers to text: "Tuesday, sunny, 28°C. Wednesday, partly cloudy, 25°C." It converts other numbers that it received into a satellite picture that appears on your screen and a bit of soothing Chinese music evocative of the calm weather pattern.

These steps may seem complex, but they are not. Your computer represents each English letter, numeral, punctuation mark, and symbol with a predetermined number. The text portion of the forecast, which in this case is made up of fifty-three symbols (including spaces) is easily represented by a list of fifty-three numbers.

The satellite image is also represented with numbers. The picture is divided into a grid of perhaps 200 rows and 200 columns, creating 40,000 tiny square cells called *pixels* (picture elements). The cells are so small that within each one the color does not change. The computer assigns three numbers to each pixel that represent the red, green, and blue components of the color in that pixel. Because there are 40,000 pixels, a list of 120,000 numbers will represent the satellite picture. The music in the message is divided up into pieces called samples—some 20,000 every second—which are converted, just like the pixels, to numbers. Compact discs work similarly, storing in little indentations on their shiny surfaces all the numbers that represent a recording.

More complex things are represented by sequences of lists of numbers. An enormous list of numbers is required to represent all the frames of still images that make up a movie—enough to fill thousands of books or the data storage capacity of several contemporary personal computers.

Besides dealing with what we hear and see, computers also handle other kinds of information. They use numbers to represent a

What Will Be

ship's heading, which is simply the degrees shown on the ship's compass, or the speed of a car, or the temperature of an engine or of the outside air in Hong Kong—any physical quantity that can be measured with devices called *sensors*. If the quantity is changing, a sequence of numbers will represent it. There are probably more than a dozen sensors inside your car measuring all sorts of things your onboard computer needs to know to adjust the engine's fuel mixture and much more.

Numbers also represent active things, "information verbs" like software programs that transform information. Computer programs are like recipes; they're made up of many instructions that tell the computer what numbers to grab, how to change them, and where to store them. The fascinating thing here is that because computer programs are written with characters and symbols, they too are represented by numbers that correspond to these symbols. This means that programs can work on other programs in order to change them. That's how a chess-playing program modifies itself as it learns from the moves, victories, and losses of its opponents.

I could go on with many more examples. But let's stop here, hoping that the first pillar is well understood: All information can be represented by numbers!

2. Two-Fingered Monsters

The second pillar of the Information Age is that machines only need the digits 1 and 0 to handle all these numbers. Unlike people, who have ten fingers (that's why we use numbers from 0 to 9) and two hands, computers have only two fingers on several million "hands." Whereas we count like this: 0, 1, 2, 3, 4, 5; they count like this: 0, 1, 10, 11, 100, 101. This smallest indivisible representation of information, this number that is either 1 or 0, is called a *bit,* from the phrase *binary digit.* In our Hong Kong forecast, each letter in the word *sunny* is represented by an eight-bit binary number, and the word is built by linking the five numbers in sequence.

A computer uses this apparently bizarre representation because it is made up of hundreds of millions of tiny transistors. A transistor is a minute device that acts like a switch. It is either closed, letting electric current pass through, or open, blocking the flow. Because a switch can only be closed or open, it can only represent one of two numbers, 1 or 0. These switches are the two-fingered hands that do the machine's counting.

Today's computers are built with *chips*—small pieces of silicon the size and thickness of your fingernail. Millions of transistors are etched onto the surface of a chip, and they are interconnected. Thus at any instant a single chip can represent several million 1s or 0s. So even our rather detailed weather map, with 120,000 pixel numbers, can fit in a portion of the transistors on one chip. Furthermore, each transistor can be turned on or off in less than one hundred millionth of a second, changing its value from 1 to 0 or vice versa.

Your personal computer may contain ten to a hundred chips. Large computers may have thousands, some devoted to storing information, others to transforming it, and yet others to transporting it. Chips are inexpensive because they are "printed" in large numbers through a complex fabrication process in a factory that costs a billion dollars to build. They can store a lot of information or perform complicated functions because of the many transistors they can hold.

A computer gets its information from input devices that transform some physical activity like keystrokes, air impulses, or temperatures to 1s and 0s. It can also record the stream of 1s and 0s coming from other computers that are wired to it or streams that are sent over a telephone line. As the numbers enter the computer, they are stored in solid-state "switches" that are set either open or closed. This constitutes the computer's *memory*. The memory is also used to store the programs that tell the machine how to process the information it gets. The switches are organized in sets of eight, called *bytes*. The number eight was chosen because a binary number with eight digits is enough to represent the characters and symbols of a keyboard.

Taken together, all the chips that hold the numbers that a computer is working on at any given time are called its *primary memory* (the *RAM* in your PC). It is fast but forgetful; when you turn the power off, the information is lost. That is why computers also have a *secondary memory* (the hard disk), which preserves the numbers even when the power is turned off. If you want to work on this data, the computer moves it to the primary memory where it can manipulate it more rapidly. When you hit Save, it copies the new data from the RAM back onto the hard disk.

Information can also be stored permanently on CD-ROMs much in the way that audio compact discs store your favorite music. *ROM* stands for *read-only memory*, which means you can retrieve data from it but you can't change the data that's there and you can't add any new data. This is just fine for electronic encyclopedias, games, photographs, and your old but precious data, information that is valuable but doesn't need to change. CD-ROMs are used because they can store 1s and 0s more compactly and inexpensively than magnetic media like hard disks. In the late-1990s the price of hard disk storage was about fifteen cents for a million bytes (megabyte), the equivalent of a paperback book. Cheap enough. CD-ROMs were even cheaper at about one cent per megabyte. These costs continue to drop.

The disadvantage of hard disks and CD-ROMs is that they are not as fast as solid-state primary memories. Thus, every computer has a balance of the two. A typical personal computer in the late-1990s had a few megabytes of primary memory and a couple of thousand megabytes (gigabytes) of hard disk memory and could read CD-ROMs that hold about 650 megabytes each.

Modern computers are roughly thirty years old. During this time, a new state-of-the-art computer chip was introduced every eighteen to twenty-four months that was roughly the same size but had twice the number of transistors on it. Once introduced commercially, its cost would be about the same too. And because the transistors were smaller, they could switch faster—an added benefit. Techies came to refer to the approximate doubling in performance every two years as *Moore's Law*, named after Gordon Moore, a founder of Intel who first observed the relationship. Moore's Law is likely to bottom out in a decade or so, but it has held up remarkably well, and it is one of the yardsticks technologists use to predict future capabilities of computers.

The venerable IBM 7094 machine that I used for my thesis in 1964 cost about $6 million of today's money. Thanks to the continuous miniaturization of transistors on chips, it has now been replaced by a $3,000 machine on my desk that is a hundred times faster—a net improvement in speed-over-cost of 200,000, or in business terms, a 20 million percent gain! It is the equivalent of improving cars of the same time period to ones today that would cost $9 and reach a top speed of 6,000 mph! No other technology has exhibited such an astonishing improvement over time. And no information revolution would be taking place today without this relentless progress.

Processing power is usually measured in how many millions of instructions a computer can perform in one second *(MIPS)*. It can also be measured by how many millions of times the fastest circuit can switch on and off in one second *(megahertz)*. MIPS and mega-hertz are used to rate computers much as maximum revolutions per minute are used to rate a car's engine.

Impressive as these terms may sound, they are poor bench-marks. They don't tell us how fast a computer can do a job we are interested in, just how fast it can spin its wheels. So when buying a computer, whether you are an individual or the CEO of a $50 bil-lion company, don't get too impressed by claims of MIPS and megahertz. Instead, try the machine on something that matters to you. The numbers will be far less impressive. No one will claim that a word processing program today is 200,000 times more effective than in the old days!

3. Computer Labor

We now come to the third pillar of the Information Age—the ap-parently miraculous way computers process information. They do it by mathematically manipulating those precious 1s and 0s.

You've read the forecast for Hong Kong. Now you call up the satellite image. When it appears on your screen, however, the cloud patterns are hard to distinguish over the faint brown land and light blue seas. You tell the computer to increase the picture's contrast. How does it comply?

The computer moves its "contrast sharpening program," which is waiting on the hard disk, to the RAM. To carry out the instruc-tions written in that program, it also locates in the RAM the three red-green-blue numbers for each pixel. It then "pushes" the num-bers down if they are below a certain threshold value, making the pixel darker, or pushes them up if they are above the threshold, making the pixel lighter. The new image will have darker dark pix-els and brighter bright pixels than the original—greater contrast.

These operations sound incredibly tedious, because we imagine with justifiable fear that we are the ones closing and opening switches and multiplying each of the 120,000 pixel values. But to a computer that loves to do the same simple things over and over at tremendous speeds, these operations are embarrassingly easy.

A typical program like a word processor requires a few million instructions to enact the program's many features like cut, paste, change font, indent, and check spelling. The term *software* is used because the instructions can be easily changed by the programmer. *Hardware,* by contrast, is very difficult to change; once tiny transis-tors are etched into silicon, the party's over. Software is very time-consuming to develop, but once a program is written it can be put into action on millions of machines. Software is the primary fuel of the Information Age, as important as the fossil fuels have been for the Industrial Age.

Every computer has one or more *processors*. They are the laborers that process information—chips that carry out the instructions given by software programs. A processor performs each instruction by shifting 1s and 0s, moving bits from transistors in one part of memory to another part, testing new bits to see if they are 1s and 0s, doing arithmetic on groups of bits, and carrying out a host of other manipulations on 1s and 0s that are neither important nor exciting in themselves. But chained together in huge sequences of millions of instructions, they perform the dazzling feats of delivering text and altering satellite maps that tell us the weather in Hong Kong.

4. Faster Than Smoke Signals

The fourth pillar of the Information Age is that information is moved around by shipping 1s and 0s across space.

Telecommunications is the reliable and rapid transportation of information from one place to another. This technology has also undergone a dramatic improvement in its cost to ship a unit of information, dropping by half every four or five years. Today, the 1s and 0s that represent information can be sent over a wire by turning an electrical current on and off rapidly. They can be sent over a glass fiber by turning a light source on and off rapidly. And they can be sent through the air by switching a radiowave transmitter between two frequencies. These technologies are called *digital* because they transfer the two digits 1 and 0 that make up the binary number system.

When your computer is ready to contact Hong Kong, it dials the weather computer's number there. Once the two computers have told each other they are ready to communicate, your computer sends a series of 1s and 0s that represent the instruction "get Hong Kong weather." The other computer returns a string of 1s and 0s that represents the text and satellite image.

Shipping 1s and 0s around is new. The older *analog* technology still dominates telephony and television. In this scheme, the current in a wire is smoothly varied up and down between, say, 0 and 9 units.

Analog systems are fine for many applications, but they are too imprecise and susceptible to fluctuations. To show this to my former sophomore classes, I would bring in a measuring cup and pour in 4.0 units of water. I would then jiggle the cup spastically, spilling some water to demonstrate how analog systems can be affected by heat or interruption from other signals. I would then ask a student to read the corrupted amount. Was it still 4.0 units, or 3.9, or 3.8? It was hard to tell. To represent the digital scheme, I would fill one glass and leave two empty, to represent the number 4 as 100. I would spill some water from the full glass, sometimes even into the empty glasses, and would ask the student to read the results. The answer would always be 100, because any amount above half-full would be read as a 1 and any amount below half-full would be read as a 0—just as it is done inside a computer. The result is exactly 4, with no uncertainty. The experiment would inevitably end with my "accidentally" pouring the water over my head—or the student's— to get the laughter and adrenal shock so essential to learning.

The problem with analog technology is that the signals get corrupted easily. Why is the audio from a compact disc so pristine compared with that of a vinyl record? Because each note is stored digitally with 1s and 0s. On a record, the vinyl grooves can warp, the needle can get worn, and dust can fill in the little pits in the grooves, all of which can make a perfectly beautiful musical sound of 4 sound scratchy or a little off, like a 3.9 or a 4.1.

Computers moved forty years ago from the analog to the digital world. Telecommunications is moving more slowly, because of the large investment in analog telephones and televisions built up over the last fifty years, but it should be mostly digital throughout the world in another decade or two.

Like computer technology, the technology of communications has benefited from the transistor and the chip. But it has also benefited from advances in glass fibers, satellites, and wireless. One very thin hair of glass fiber has the surprising ability to carry information very rapidly across thousands of miles in the form of light pulses that turn on and off as fast as 10 billion times per second. A two-hour movie can be sent through a glass fiber in a few seconds, while it would take a month through an ordinary phone line. The speed of signals through different kinds of pipes is a crucial factor in determining the success of the Information Marketplace.

The speed limits on any method of communication can be better understood if we visualize a simple model, like the sending of smoke signals. If we cover and uncover a fire with our blanket at increasing speed, there will come a point at which the puffs of smoke begin to blend into each other, conveying nothing. The highest rate at which we can flap our blanket and still keep a separation between the puffs of smoke is called the system's *bandwidth*. The limit is about two flaps per second—enough to communicate a 1 or 0 every second, in other words, one bit per second. Sending the fifty-three characters in the text of our Hong Kong forecast in this smoky way would take about seven minutes. A two-hour movie would take a mere half-century!

Phone lines today are 60,000 times faster than smoke signals. Coaxial cable and satellite links are about 500,000 times faster. The best experimental glass fibers are 1 trillion times faster.

Bandwidth limitations can be overcome in part with a technical trick called *compression*. It requires that the sender and receiver agree on various shorthands, for example, *C* for cloudy, *S* for sunny, *R* for rain. When a weather program sends out a forecast, it can simply send the letters instead of the words, reducing the amount of data that must be sent and making the sending of the forecast faster—as if the pipe had more bandwidth to transmit the package of information. If a large, rectangular portion of the sky in a picture is all blue, instead of sending 6,000 bits to represent the pixels in this part of the image, a compression program might send 60 bits that define the color once and the rectangular area that should be filled with that color. The compression in this case is 100 to 1, which is great, because the same information can be transmitted 100 times faster. Of course, other more detailed parts of the picture will not be as compressible. Today's schemes for television and computer images achieve an overall average compression of 10 to 50, still greatly improving transmission speed. Compression factors like this make possible the transport of a video over a pair of phone wires or the squeezing of 1,000 video channels into one coaxial cable.

5. Computer Networks

The first three pillars make up the technology of computers; the fourth pillar is the technology of communications. When these two queen technologies of computers and communications come together, they give rise to one additional pillar: *networking*. It is the foundation of all information infrastructures.

As soon as you connect one computer to a second computer, you have a *network*. If a bunch of computers are linked within a small area like a building, they are known as a local area network *(LAN)*.

If they are connected across a university campus, a large corporate compound, or a few square miles of users, they are called wide area networks *(WANs)*. These names are approximate, and there is mild debate as to their precise definitions.

Your computer is connected to a wire, telephone line, or glass fiber through an *interface*—a little box sitting next to your computer or a circuit board inside it. If it connects to the telephone line, it is called a *modem*. Alternately, it may connect your computer to the television cable in your home (video modem) or to the airwaves of a cellular, wireless, or satellite system.

Most organizations want to link their *intranets* (internal networks) to the greater Information Marketplace networks outside their confines so that they may belong to and take advantage of the rich world of information out there. But they also fear that this interconnection may compromise the confidentiality of their or their clients' dealings. What they need in order to accomplish both objectives is a *firewall*—a way of controlling what information can and cannot pass between the internal and external networks. This is usually done with one computer linking the internal and the external networks. All inside-outside traffic passes through this carefully programmed machine—the firewall—where it is examined and controlled. Firewalls cannot be 100 percent secure, because information that looks benign to the machines could be allowed to pass through even though it contains a hidden, sinister program. Once inside the firewall, like the old Trojan Horse, the program opens up and lets out its warriors to do their ugly deeds—steal data, copy passwords, or grab secret information.

Researchers are constantly fooling around with different network flavors, like experimental "desk area networks" that would hook together all the gadgets on your desk.

Regardless of their name, networks accomplish the same goal—transporting information rapidly and reliably among geographically dispersed computers. The purpose of networks is similar to that of the postal system—to ship information to the right destination. And like the postal system, networks must deliver the information intact, rapidly, and ideally without allowing anyone to know the contents of what is being transported.

Networks route their information packages in a way that is also similar to the postal system. Your interface sends your computer's information package to a local collection center known as an intermediate network node. From there it is routed through other regional or national nodes to reach its destination. Like the postal system, a good network will keep track of nodes that have failed and will route the packages around them through other nodes so they will reach the proper recipient.

The analogy breaks down in the way information is packaged, however. Computers do not pack all the information they want to send into a single *packet*. Instead, they use many small packets, each consisting of a few thousand bytes of 1s and 0s. In postal terms, this would be equivalent to sending each page of a five-page letter in five envelopes. One important reason computers do this is that, at their scale of doing things, the information they send is often voluminous. If it were sent in one packet, it would take a long time for the system to move the bulky item. And if some information were lost in the process, the entire huge packet would have to be resent. By using many smaller packets, networks can retransmit only the packets that become corrupted.

Like the post office, a network needs the name and address of the recipient to figure out where to route the packets. A typical Internet name might read: jr@csl.mit.edu. The initials *jr* would correspond to a person at (@) a place called *csl* that is part of a group called *mit* that belongs to a logical group made up of educational institutions, called *edu*. This division of names into domains, as they are called, is the main logical organizational principle of the Internet naming scheme. Special computers called *domain name servers* (DNS) convert these names to numbers that stand for physical Internet addresses.

As the scheme is currently used, it favors a spatial organization. A packet for *jr* will be sent to the computer on her desk. If she's at home, in her car, or off to Hong Kong, she won't get it. In principle, this naming scheme could be used to let jr redirect her location to any one of these places. This will probably come about.

The Web is nothing more than a specific way of using these addressing and transport capabilities of the Internet. By adhering to a set of agreements called *protocols*, home pages, browsers and other software make possible the familiar point-click-and-follow-the-link behavior that has made the Web famous. The primary agreements are the *universal resource locator* (url), which establishes a unique address for any and every piece of information made available on the Internet—a document, picture, sound, or video snippet; the *hypertext markup language* (html)—a uniform way of representing information like text headings, paragraphs, images, and sounds on any computer system; and a set of conventions called *hypertext transport protocol* (http) for linking and transporting this information locally or across vast distances, with equal ease.

Computers and communications are the two queen technologies of the Information Age. Combined into computer networks, they offer the basis for today's Internet and Web and for tomorrow's information infrastructures. Hence they are the technological foundation of the Information Marketplace. The great capabilities they offer are all a consequence of the five pillars: numbers can represent all information; these numbers are made up of 1s and 0s; computers transform information by doing arithmetic on these 1s and 0s; information is communicated by shipping 1s and 0s across space; and computer networks are formed by combining computers and communications.

Index

18–20; micro-cell wireless, 49–50; privacy of computer, 98–99, 102; satellite, 48–49. *See also* telecommunications

compression, 357

CompuServe, 33, 289

The Computer Age: A Twenty-Year View (Dertouzos and Moses), 128

computer-aided design (CAD), 268

computer community: shared games of, 29–30; of time-shared system, 27–28. *See also* virtual neighborhoods

computer companies, 44

computer crime: the first, 27–28; types of, 286–89; undetected, 28–29

"computer English," 86

computer languages: as automatization tool, 84–87; C++, 263; Java and Fortran, 84; Logo, 176–77

computer networks, 357–60

computer programming, 264–67

computers: communication between, 83–89; current information infrastructure and, 18–20; faults listed for, 254–62; interfaces for talking to, 56–64; labor of, 354–55; Moore's Law on, 134, 353; network, 109; personal, 31–35; processing of, 351–54; programming for, 264–67; time-shared, 27–28; user friendly, 262–63

computer security, 98–107

Corbató, Fernando, 27

crime (computer), 27–29, 286–89

cryptography, 99–103, 221–22, 320

cultural veneer, 283–84

culture, attacks on, 292

Cyberspace, 22

Defense Advanced Research Projects Agency (DARPA), 7, 35–36

data sockets, 115–17

Davos World Economic Forum (1995), 283–84

Delors, Jacques, 7, 284

democratization of art, 154–55

Dennet, Daniel, 72

derived demand goods/services, 236

DES (Digital Encryption Standard), 100–101, 103

design tools, 182–83

Diffie, Whitman, 101

digital cameras, 149

digital cash, 105

Digital Orrery, 203

digital signatures, 103

digital technologies, 355–56

digital videos, 149–50

direct electronic commerce, 193

discussion groups (Internet), 136–38

di Sessa, Andy, 177

DivorceNet, 201

DNA-based defense, 220

domain name servers (DNS), 359

domains, 359

Dorsey, Julie, 133

Dow Jones, 33

Edifact standards, 194

education: Information Marketplace and, 178–86; roles of schools/community in, 187–89; through virtual reality, 130–34

edutainment CD-ROMs, 179

egalitarianism, 211

electronic bulldozers, 229, 252–53, 270–71, 303–4, 328–29

electronic cash, 105

electronic checks, 104

electronic commerce, 193–96

electronic forms, 85

Electronic Mail and Message Systems: AFIPS, 9

electronic nose interface, 73–74

electronic polling, 216

electronic proximity: assessing effects of, 229, 278–80, 329; Big Brother and, 293–94; censorship and, 289–91; described, 277, 304–5; fears associated with, 277–78; impact on crime/law of, 286–89; universal culture and, 282–86; urban villager and, 280–81

electronic town hall meetings, 217

electronic voting, 216

electronic White and Yellow Pages, 34

e-mail: addresses, 359; of